PRINCIPLES

OF

TRANSFORMER DESIGN

BY

ALFRED STILL

M.Inst.C.E., Fel.A.I.E.E., M.I.E.E.

Professor of Electrical Design, Purdue University,
Author of "Polyphase Currents,"
"Electric Power Transmission," etc.

FIRST EDITION

NEW YORK:

JOHN WILEY & SONS, Inc.

London: CHAPMAN & HALL, Limited

1919

PRESS OF
BRAUNWORTH & CO.
BOOK MANUFACTURERS
BROOKLYN, N. Y.

PREFACE

A BOOK which deals exclusively with the theory and design of alternating current transformers is not likely to meet the requirements of a College text to the same extent as if its scope were broadened to include other types of electrical machinery. On the other hand, the fact that there may be a limited demand for it by college students taking advanced courses in electrical engineering has led the writer to follow the method of presentation which he has found successful in teaching electrical design to senior students in the school of Electrical Engineering at Purdue University. Stress is laid on the fundamental principles of electrical engineering, and an attempt is made to explain the reasons underlying all statements and formulas, even when this involves the introduction of additional material which might be omitted if the needs of the practical designer were alone to be considered.

A large portion of Chapter II has already appeared in the form of articles contributed by the writer to the *Electrical World*; but the greater part of the material in this book has not previously appeared in print.

LaFayette, Ind.
January, 1919

CONTENTS

CHAPTER I

ELEMENTARY THEORY.—TYPES.—CONSTRUCTION

CHAPTER II

INSULATION OF HIGH-PRESSURE TRANSFORMERS

v

CHAPTER III

EFFICIENCY AND HEATING OF TRANSFORMERS

CHAPTER IV

MAGNETIC LEAKAGE IN TRANSFORMERS.—REACTANCE.—REGULATION

CHAPTER V

PROCEDURE IN TRANSFORMER DESIGN

CHAPTER VI

Transformers for Special Purposes

LIST OF SYMBOLS

A = area of equipotential surface perpendicular to lines of force (sq. cm.).

A = cross-section of iron in plane perpendicular to laminations (sq. in.).

a = ampere-turns per inch length of magnetic path.

a = total thickness of copper per inch of coil measured perpendicularly to layers.

B = magnetic flux per sq. cm. (gauss).

B_{am} is defined in Art. 9.

b = total thickness of copper per inch of coil measured through insulation parallel with layers.

$b = \dfrac{W_c}{W_t}.$

C = electrostatic capacity; or permittance,

$= \dfrac{\text{coulombs}}{\text{volts}}$ = flux per unit e.m.f. (farad).

C_{mf} = capacity in microfarads.

c = a coefficient used in determining V_t.

D = flux density in electrostatic field $= \dfrac{\Psi}{A} = KkG$ (coulombs per sq. cm.).

E = e.m.f. (volts), usually r.m.s. value, but sometimes used for max. value.

E_1 = virtual value of induced volts in primary $\left(= E_2 \times \dfrac{T_p}{T_s}\right)$.

E'_1 = component of impressed voltage to balance E_1.

E_2 = secondary e.m.f. produced by flux Φ; induced secondary e.m.f.

E_e = primary voltage equivalent to secondary terminal voltage $\left(= E_s \times \dfrac{T_p}{T_s}\right)$.

E_p = e.m.f. (volts) applied at primary terminals.

E_s = secondary terminal voltage.

E_2 = impressed primary voltage when secondary is short-circuited.

e = e.m.f. (volts).

F = force (dynes).

f = frequency (cycles per second).

$G = \dfrac{de}{dl}$ = potential gradient (volts per centimeter).

g = distance between copper of adjacent primary and secondary coils, in centimeters (Fig. 42).

H = magnetizing force, or m.m.f. per cm.

h = length (cms.) defined in text (Fig. 42).

I = r.m.s. value of current (amps.).

I_1 = balancing component of primary current $= I_s \left(\dfrac{T_s}{T_p}\right)$.

I_c = current in the portion of an auto-transformer winding common to both primary and secondary circuits.

I_e = total primary exciting current.

I_0 = " wattless " component of I_e (magnetizing component).

I_p = total primary current.

I_s = total secondary current.

I_w = " energy " component of I_e (" in-phase " component).

$K = 8.84 \times 10^{-14}$ farads per cm. cube = the specific capacity of air.

$\left.\begin{array}{l} K_r \\ K_s \end{array}\right\}$ definition follows formula (34) in Art. 27.

Kv = kilovolts.

k = dielectric constant or relative specific capacity, or permittivity ($k = 1$ for air).

k = heat conductivity (watts per inch cube per 1° C.).

k = coefficient used in calculating the effective cooling surface of corrugated tanks.

k_c = about 1.8×10^{-6} for copper.

k_t = (refer text (Art. 39) for definition).

l = length (cms.).

l = mean length, in centimeters, of projecting end of transformer coil.

l = length measured along line or tube of induction (cms.).

l_c = mean length per turn of windings.

l_t = mean length of magnetic circuit measured along flux lines.

M_c = weight of copper in transformer coils (lbs.)

M_0 = weight of oil in transformer tank (lbs.).

$m = 2\pi f \times 10^{-8}$.

$n = \dfrac{l}{\lambda}$ (in formula for calculating cooling surface of corrugated tanks).

n = usually from 1.6 to 2 in B^n (core loss formulas).

P = weight of iron in transformer core (or portion of core), lbs.

p = thickness of half primary coil in centimeters (defined in text in connection with Fig. 42).

R = resistance (ohms).

R_1 = resistance of primary winding (ohms).

R_2 = resistance of secondary winding (ohms).

R_h = " thermal ohms."

R_p = equivalent primary resistance = $R_1 + R_2 \left(\dfrac{T_p}{T_s}\right)^2$.

r = ratio $\dfrac{\text{total number of turns}}{\text{number of turns common to both circuits}}$ (auto-transformers).

S = effective cooling surface of transformer tank (sq. in).

s = thickness of half secondary coil (cms.) defined in text (Fig. 42).

T = number of turns in coil of wire.

T_1 = number of turns in half primary group of coils adjacent to secondary coil.

T_2 = number of turns in half secondary group of coils adjacent to primary coil.

T_d = difference of temperature (degrees centigrade).

T_0 = initial oil temperature.

T_p = number of turns in primary winding.

T_s = number of turns in secondary winding.

T_t = oil temperature at end of time t_m minutes.

t = thickness (usually inches).

t = interval of time (seconds).

t_m = interval of time (minutes).

V_t = volts induced per turn of transformer winding.

W = power (watts).

W_c = full-load copper loss (watts).

W_i = core loss (watts).

W_t = total transformer losses (watts).

w = watts dissipated per sq. in. of (effective) tank surface.

w = watts lost per lb. of iron in (laminated) core.

X_1 = reactance (ohms) of one high-low section of winding.

X_ρ = reactance (ohms) commonly referred to as equivalent **primary** reactance.

Z_ρ = impedance (ohms) on short circuit.

θ = phase angle (cos θ = power factor of external circuit).

θ = "electrical" angle (radians) = $2\pi ft$.

λ = pitch of corrugations on tank surface.

Φ = magnetic flux (Maxwells) in iron core.

ϕ = phase angle (cos ϕ = power factor on primary side of transformer).

Ψ = dielectric flux, or quantity of electricity, or electrostatic induction = CE = AD coulombs.

PRINCIPLES

OF

TRANSFORMER DESIGN

CHAPTER I

ELEMENTARY THEORY—TYPES—CONSTRUCTION

1. Introductory. The design of a small lighting transformer for use on circuits up to 2200 volts, or even 6600 volts, is a very simple matter. The items of importance to the designer are:

(1) The iron and copper losses; efficiency, and temperature rise;

(2) The voltage regulation, which depends mainly upon the magnetic leakage, and therefore upon the arrangement of the primary and secondary coils;

(3) Economical considerations, including manufacturing cost.

With the higher voltages and larger units, not only does the question of adequate cooling become of greater importance; but other factors are introduced which call for considerable knowledge and skill on the part of the designer. The problems of insulation and protection against abnormal high-frequency surges in the external circuit are perhaps the most important; but with the increasing amount of power dealt with by some modern units, the mechanical forces exerted by the magnetic flux on short-circuits, or heavy over loads, may be enormous, requiring special means of clamping or bracing the coils, to prevent deformation and damage to insulation.

Since we are concerned mainly with a study of the transformer from the view point of the designer, little will be said concerning the operation of transformers, or the advantages and disadvantages of the different methods of connecting the units on polyphase systems. It will, however, be necessary to discuss the theory underlying the action of all static transformers, and it is proposed to take up the various aspects of the subject in the following order:

Elementary theory, omitting all considerations likely to obscure the fundamental principles; brief description of leading types and methods of manufacture; problems connected with insulation; losses, heating, and efficiency; advanced theory, including study of magnetic leakage and voltage regulation; procedure in design; numerical examples of design; reference to special types of transformers.

2. Elementary Theory of Transformer. A single-phase alternating current transformer consists essentially of a core of laminated iron upon which are wound two distinct sets of coils, known as the primary and secondary windings, respectively, all as shown diagrammatically in Fig. 1.

When an alternating e.m.f. of E_p volts is applied to the terminals of the primary (P), this will set up a certain flux (Φ) of alternating magnetism in the iron core, and this flux will, in turn, induce a counter e.m.f. of self-induction in the primary winding; the action being similar to what occurs in any highly inductive coil or winding. Moreover, since the secondary coils—

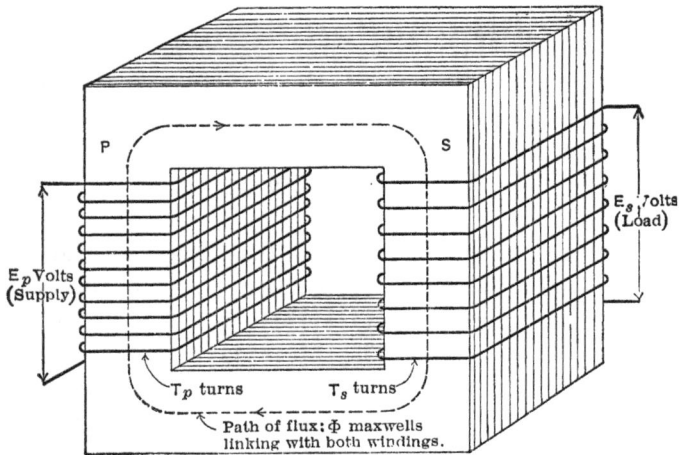

FIG. 1.—Essential Parts of Single-phase Transformer.

although not in electrical connection with the primary—are wound on the same iron core, the variations of magnetic flux which induce the counter e.m.f. in the primary coils will, at the same time, generate an e.m.f. in the secondary winding.

The path of the magnetic lines is usually through a closed iron circuit of low reluctance, in order that the exciting ampere-turns shall be small. There will always be some flux set up by the primary which does not link with the secondary, but the amount of this leakage flux is usually very small, and in any case it is proposed to ignore it entirely in this preliminary study. In this connection it may be pointed out that the design indicated in Fig. 1, with a large space for leakage flux between the primary and secondary coils, would be unsatisfactory in practice; but the assumption will now be made that the whole of the flux (Φ maxwells) which passes through the primary coils, links also with all the secondary coils. In other words, the e.m.f. induced in the winding *per turn of wire* will be the same in the secondary as in the primary coils.

Suppose, in the first place, that the two ends of the primary winding are connected to constant pressure mains, and that no current is taken from the secondary terminals. The total flux of Φ maxwells increases twice from zero to its maximum value, and decreases twice from its maximum to zero value, in the time of one complete period. The flux cut per second is therefore $4\Phi f$, and the *average* value of the induced e.m.f. in the primary is,

$$E_{average} = \frac{4\Phi f T_p}{10^8} \text{ volts,}$$

where T_p stands for the number of turns in the primary winding.

If we assume the flux variations to be sinusoidal, the

form factor is 1.11, and the virtual value of the induced primary volts will be,

$$E_1 = \frac{4.44 f \Phi T_p}{10^8}. \quad \ldots \quad \text{(1)}$$

The vector diagram corresponding to these conditions has been drawn in Fig. 2. Here OB represents the phase of the flux which is set up by the current OI_e in

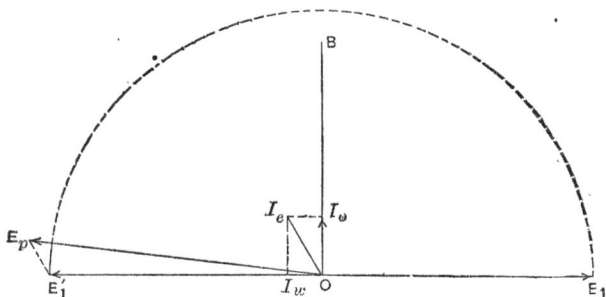

FIG. 2.—Vector Diagram of Unloaded Transformer.

the primary. This total primary exciting current can be thought of as consisting of two components: the "wattless" component OI_0 which is the true magnetizing current, in phase with the flux; and OI_w (which owes its existence to hysteresis and eddy current losses) exactly 90° in advance of the flux. The volts induced in the primary are OE_1 drawn 90° behind OB to represent the lag of a quarter period. The voltage that must be impressed at the terminals of the primary is OE_p made up of the component OE'_1 exactly equal but opposite to OE_1, and $E'_1 E_p$ drawn parallel to OI_e and representing the IR drop in the primary circuit. The actual magnitude of this component would be I_eR_1 where R_1 is the ohmic resistance of the primary; but in practice this ohmic drop is usually so small as to be negligible, and the impressed voltage E_p is virtually the same as E'_1, *i.e.*, equal in amount, but opposite in phase to the induced voltage E_1.

For preliminary calculations it is, therefore, usually permissible to substitute the terminal voltage for the induced voltage, and write for formula (1)

$$E_p = \frac{4.44 f \Phi T_p}{10^8} \text{ (approximately). . . (1a)}$$

Similarly,

$$E_s = \frac{4.44 f \Phi T_s}{10^8} \text{ (approximately), . . (1b)}$$

where E_s and T_s stand respectively for the secondary terminal voltage and the number of turns in secondary. It follows that,

$$\frac{E_p}{E_s} = \frac{T_p}{T_s}, \quad . \quad . \quad . \quad . \quad . \quad . \quad (2)$$

which is approximately true in all well-designed static transformers when no current, or only a very small current, is taken from the secondary.

3. Effect of Closing the Secondary Circuit. When considering the action of a transformer with loaded secondary, that is to say, with current taken from the secondary terminals, it is necessary to bear in mind that —except for the small voltage drop due to ohmic resistance of the primary winding—the counter e.m.f. induced by the alternating magnetic flux in the core must still be such as to balance the e.m.f. impressed at primary terminals. It follows that, with constant line voltage, the flux Φ has very nearly the same value at full load as at no load. The m.m.f. due to the current in the secondary windings would entirely alter the magnetization of the core if it were not immediately counteracted by a current component in the primary windings of exactly the same magnetizing effect, but tending at every instant to set up flux in the *opposite* direction. Thus, in order to maintain the flux necessary to produce the required counter e.m.f. in the primary, any tendency on the part of the secondary current to alter this flux is met by a flow of current in the primary circuit; and since, in well-designed transformers, the magnetizing current is always a small percentage of the full-load current, it follows that the relation

$$I_p T_p = I_s T_s, \quad . \quad . \quad . \quad . \quad . \quad (3)$$

is approximately correct.

Thus,
$$\frac{I_p}{I_s} = \frac{T_s}{T_p},$$

where I_p and I_s stand respectively for the total primary and secondary current.

The open-circuit conditions are represented in Fig. 3 where E_p is the curve of primary impressed e.m.f. and I_e is the magnetizing current, distorted by the hysteresis of the iron core, as will be explained later. E_s is the curve of secondary e.m.f. which coincides in phase with the primary induced e.m.f. and is therefore—if we neglect the small voltage drop due to ohmic resistance of the primary—exactly in opposition to the impressed e.m.f. The curve of magnetization (not shown) would

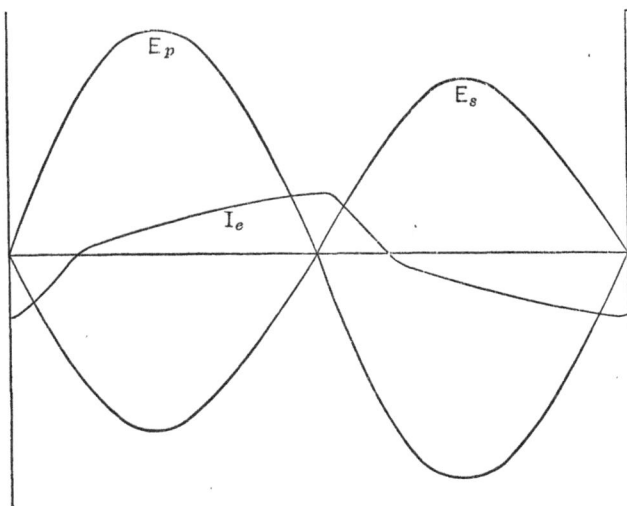

FIG. 3.—Voltage and Current Curves of Transformer with Open Secondary Circuit.

be exactly a quarter period in advance of the induced, or secondary, e.m.f.

In Fig. 4, the secondary circuit is supposed to be closed on a non-inductive load, and the secondary current, I_s will, therefore, be in phase with the secondary e.m.f.

The tendency of the secondary current being to produce a change in the magnetization of the core, the current in the primary will immediately adjust itself so as
to maintain the same (or nearly the same) cycle of magnetization as on open circuit; that is to say, the flux

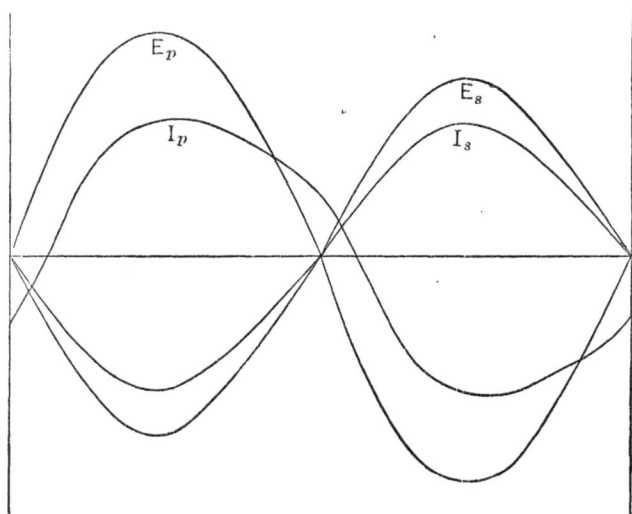

FIG. 4.—Voltage and Current Curves of Transformer on Non-inductive
Load.

will continue to be such as will produce an e.m.f. in the
primary windings equal, but opposite, to the primary
impressed potential difference. The new curve of primary current, I_p (Fig. 4), is therefore obtained by adding
the ordinates of the current curve of Fig. 3 to those of
another curve exactly opposite in phase to the secondary

current, and of such a value as to produce an equal magnetizing effect.

4. Vector Diagrams of Loaded Transformer Without Leakage. The diagram of a transformer with secondary closed on a non-inductive load is shown in Fig. 5. In order to have a diagram of the simplest kind, not only the leakage flux, but also the resistance of the windings

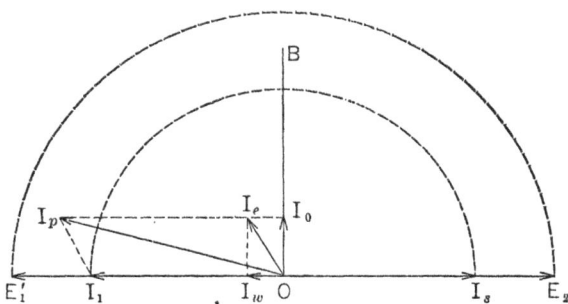

FIG. 5.—Vector Diagram of Transformer on Non-inductive Load.

will be considered negligible. The vectors then have the following meaning:

OB = Phase of flux Φ linked with both primary and secondary windings;

I_e = Exciting current necessary to produce flux Φ;

E_2 = Secondary e.m.f. produced by alternations of the flux Φ;

E'_1 = Primary e.m.f. equal, but opposite, to the e.m.f. produced by alternations of the flux Φ (In this case it is equal to the applied e.m.f., since the IR drop is negligible);

I_s = Current drawn from secondary; in phase with E_2;
I_1 = Balancing component of primary current, drawn
exactly opposite to I_s and of value $I_s \times \dfrac{T_s}{T_p}$;
I_p = Total primary current, obtained by combining
I_1 with I_e.

In Fig. 6 the vectors have the same meaning as above, but the load is supposed to be partly inductive, which accounts for the lag of I_s behind E_2.

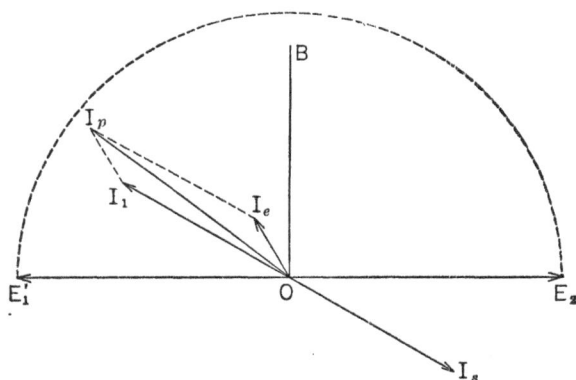

FIG. 6.—Vector Diagram of Transformer on Inductive Load.

It is convenient in vector diagrams representing both primary and secondary quantities to assume a 1 : 1 ratio in order that balancing vectors may be drawn of equal length. The voltage vectors may, if preferred, be considered as *volts per turn*, while the secondary current vector can be expressed in terms of the primary current by multiplying the quantity representing the actual secondary current by the ratio $\dfrac{T_s}{T_p}$.

5. Polyphase Transformers. Although we have considered only the single-phase transformer, all that has been said applies also to the polyphase transformer because each limb can be considered separately and treated as if it were an independent single-phase transformer.

In practice it is not unusual to use single-phase transformers on polyphase systems, especially when the units are of very large size. Thus, in the case of a three-phase transmission, suppose it is desired to step up from 6600 volts to 100,000 volts, three separate single-phase transformers can be used, with windings grouped either Y or Δ, and the grouping on the secondary side need not necessarily be the same as on the primary side. A saving in weight and first cost may be effected by combining the magnetic circuits of the three transformers into one. There would then be three laminated cores each wound with primary and secondary coils and joined together magnetically by suitable laminated yokes; but since each core can act as a return circuit for the flux in the other two cores, a saving in the total weight of iron can be effected. Except for the material in the yokes, this saving is similar to the saving of copper in a three-phase transmission line using three conductors only (as usual) instead of six, as would be necessary if the three single-phase circuits were kept separate. In the case of a two-phase transformer, the windings would be on two limbs, and the common limb for the return flux need only be of sufficient section to carry $\sqrt{2}$ times the flux in any one of the wound limbs.

It is not always desirable to effect a saving in first

cost by installing polyphase tiansformers in place of
single-phase units, especially in the large sizes, because,
apart from the increased weight and difficulty in hand-
ling the polyphase transformer, the use of single-phase
units sometimes leads to a saving in the cost of spares to
be carried in connection with an important power development. It is unusual for all the circuits of a polyphase
system to break down simultaneously, and one spare
single-phase transformer might be sufficient to prevent
a serious stoppage, while the repair of a large polyphase
transformer is necessarily a big undertaking.

6. Problems in Design. The volt-ampere input of a
single-phase transformer is $E_p I_p$, and if we substitute
for E_p the value given by formula ($1a$), we have

$$\text{Volt-amperes} = \frac{4.44 f}{10^8} \times \Phi \times T_p I_p.$$

Thus, for a given flux Φ, which will determine the cross-
section of the iron core, there is a definite number of
ampere turns which will determine the cross-section of
the winding space. There is no limit to the number of
designs which will satisfy the requirements apart from
questions of heating and efficiency; but there is obvi-
ously a relation between the weight of iron and weight
of copper which will produce the most economical
design, and this point will be taken up when discussing
procedure in design. It will, however, be necessary to
consider, in the first place, a few practical points in
connection with the construction of transformers, and
also the effect of insulation on the space available for the
copper. The predetermination of the losses in both iron

and copper must then be studied with a view to calculating the temperature rise and efficiency. Finally, the flux leakage must be determined with a reasonable degree of accuracy because this, together with the ohmic resistance of the windings, will influence the voltage regulation, which must usually be kept within specified limits.

7. Classification of Alternating-current Transformers. Since we are mainly concerned with so-called constant-potential transformers as used on power and lighting circuits, we shall not at present consider constant-current transformers as used on series lighting systems and in connection with current-measuring instruments; neither shall we discuss in this place the various modifications of the normal type of transformer which render it available for many special purposes.

Transformers might be classified according to the method of cooling, or according to the voltage at the terminals, or, again, according to the number of phases of the system on which they will have to operate.

Methods of cooling will be referred to again later when treating of losses and temperature rise; but, briefly stated, they include:

(1) Natural cooling by air.

(2) Self-cooling by oil; whereby the natural circulation of the oil in which the transformer is immersed carries the heat to the sides of the containing tank.

(3) Cooling by water circulation: a method generally similar to (2) except that coils of pipe carrying running water are placed near the top of the tank below the surface of the oil.

(4) Cooling with forced circulation of oil: a method used sometimes when cooling water is not available. It permits of the oil being passed through external pipe coils having a considerable heat-radiating surface.

(5) Cooling by air blast; whereby a continuous stream of cold air is passed over the heated surfaces, exactly as in the case of large turbo-generators.

In regard to difference of voltage, this is mainly a matter of insulation, which will be taken up in Chap. II. The essential features of a potential transformer are the same whether the potential difference at terminals is large or small, but the high-pressure transformer will necessarily occupy considerably more space than a low-pressure transformer of the same k.v.a. output. The difficulties of avoiding excessive flux leakage and consequent bad voltage regulation are increased with the higher voltages.

Low-voltage transformers are used for welding metals and for any purpose where very large currents are necessary, as for instance, in thawing out frozen water pipes, while transformers for the highest pressures are used for testing insulation. Testing-transformers to give up to 500,000 volts at secondary terminals are not uncommon, while one transformer (at the Panama-Pacific Exposition of 1915) was designed for an output of 1000 k.v.a. at 1,000,000 volts. This transformer weighed 32,000 lb., and 225 bbl. of oil were required to fill the tank in which it was immersed.

A classification of transformers by the number of phases would practically resolve itself—so far as present-day tendencies are concerned—into a division between single-phase and three-phase transformers. From the point of view of the designer, it will be better to consider the use to which the transformer—whether single-phase or polyphase—will be put. This leads to the two classes:

(1) Power transformers.

(2) Distributing transformers.   .

Power Transformers. This term is here used to include all transformers of large size as used in central generating stations and sub-stations for transforming the voltage at each end of a power transmission line. They may be designed for maximum efficiency at full load, because they are usually arranged in banks, and can be thrown in parallel with other units or disconnected at will. Artificially cooled transformers of the air-blast type are easily built in single units for outputs of 3000 k.v.a. single-phase and 6000 k.v.a. three-phase; but the terminal pressure of these transformers rarely exceeds 33,000 volts. A three-phase unit of the air-blast type with 14,000 volts on the high-tension windings has actually been built for an output of 20,000 k.v.a. For higher voltages the oil insulation is used, generally with water cooling-pipes. These transformers have been built three-phase up to 10,000 k.v.a. output from a single unit, for use on transmission systems up to 150,000 volts.* With the modern demand for larger

* The 10,000 k.v.a. three-phase, 6600 to 110,000-volt units in the power houses of the Tennessee Power Company on the Ocoee River weigh about 200,000 lb.; they are 19 ft. high, and occupy a floor space 20 ft. by 8 ft.

Single-phase, oil-insulated, water-cooled transformers for a frequency of 60 cycles and a ratio of 13,200 to 150,000 volts have been built for an output of 14,000 k.v.a. from a single unit.

transformers to operate out of doors, power transformers of the oil-immersed self-cooling type (without water coils) are now being constructed in increasing number. A self-cooling 25-cycle transformer for 8000 k.v.a. output has actually been built: a number of special tube-type radiators connected by pipes to the main oil tank are provided; the total cooling surface in contact with the air being about 7000 sq. ft.

Distributing Transformers. These are always of the self-cooling type, and almost invariably oil-immersed. They include the smaller sizes for outputs of 1 to 3 k.w. such as are commonly mounted on pole tops. These transformers are rarely wound for pressures exceeding 13,000 volts, the most common primary voltage being 2200.

In the design of distributing transformers, it is necessary to bear in mind that since they are continuously on the circuit, the " all-day " losses—which consist largely of hysteresis and eddy-current losses in the iron—must be kept as small as possible. In other words, it is not always desirable to have the highest efficiency at full load.

8. Types of Transformers. Construction. All transformers consist of a magnetic circuit of laminated iron with which the electric circuits (primary and secondary) are linked. A distinction is usually made between *core-type* and *shell-type* transformers. Single-phase transformers of the core- and shell-types are illustrated by Figs. 7 and 8, respectively. The former shows a closed laminated iron circuit two limbs of which carry the windings. Each limb is wound with both primary and secondary circuits in order to reduce the magnetic leakage which would otherwise be excessive. The coils may be cylindrical in form and placed one inside the other with the necessary insulation between them, or the windings may be " sandwiched," in which case flat rectangular or circular coils, alternately primary and sec-

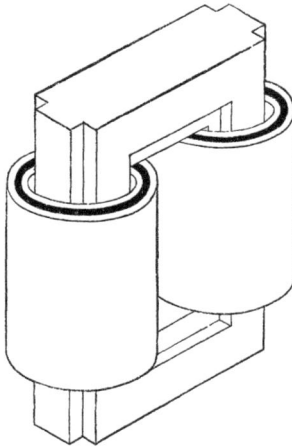

FIG. 7.—Core-type Transformer.   FIG. 8.—Shell-type Transformer.

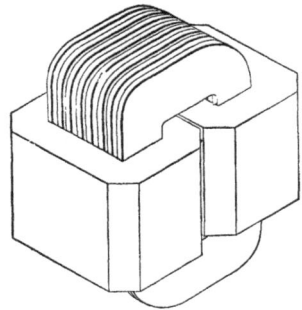

ondary, are stacked one above the other with the requisite insulation between.

Fig. 8 shows a single set of windings on a central laminated core which divides after passing through the coils and forms what may be thought of as a shell of iron around the copper. The manner in which the core is usually built up in a large shell-type transformer is shown in Fig. 9. The thickness of the laminations varies between 0.012 and 0.018 in., the thicker plates being permissible when the frequency is low. A very usual thickness for transformers working on 25- and 60-cycle circuits is 0.014 in. The arrangement of the stampings is reversed in every layer in order to cover the joints and so reduce the magnetizing component of the primary current. A very thin coating of varnish

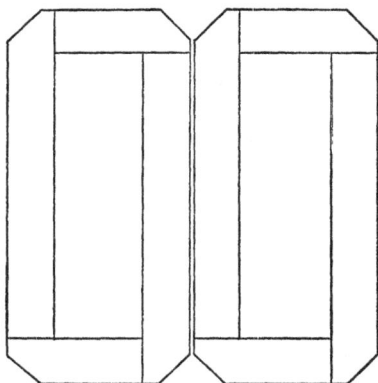

FIG. 9.—Method of Assembling Stampings in Shell-type Transformer.

or paper is sufficient to afford adequate insulation between stampings. Ordinary iron of good magnetic quality may be used for transformers on the lower frequencies, but it is customary to use special alloyed iron for 60-cycle transformers. This material has a high electrical resistance and, therefore, a small eddy-current loss. The loss through hysteresis is also small, but the permeability of alloyed iron is lower than that of ordinary iron and this tends to increase the magnetizing current. The cost of alloyed iron is appreciably higher than that of ordinary transformer iron.

The choice of type—whether "core" or "shell"—will not greatly affect the efficiency or cost of the transformer. As a general rule, the core type of construction has advantages in the case of high-voltage transformers of small output, while the shell type is best adapted for low-voltage transformers of large output.

Fig. 10 illustrates a good practical design of shell-type transformer in which a saving of material is effected by arranging the magnetic circuit to surround all four sides of a square coil. The dimensions of the iron circuit, as indicated on the sketch, show a cross-section of the magnetic circuit outside the coils exactly double the cross-section inside the coils. This will be found to lead to slightly higher efficiency, for the same cost of material, than if the section were the same inside and outside the coil. It is generally advantageous to use higher flux densities in the iron upon which the coils are wound than in the remainder of the magnetic circuit, because the increased iron loss is compensated for by the reduced copper loss due to the shorter average length per turn of the windings.

Fig. 11 illustrates a similar design of shell-type transformer in which the magnetic circuit is still further divided, and the windings are in the form of cylindrical coils. The relative positions of primary and secondary coils need not be as shown in Figs. 10 and 11, as they can be of the "pancake" shape of no great thickness, with primary and secondary coils alternating. A proper arrangement of the coils is a matter of great importance

when it is desired to have as small a voltage drop as
possible under load; but this point will be taken up

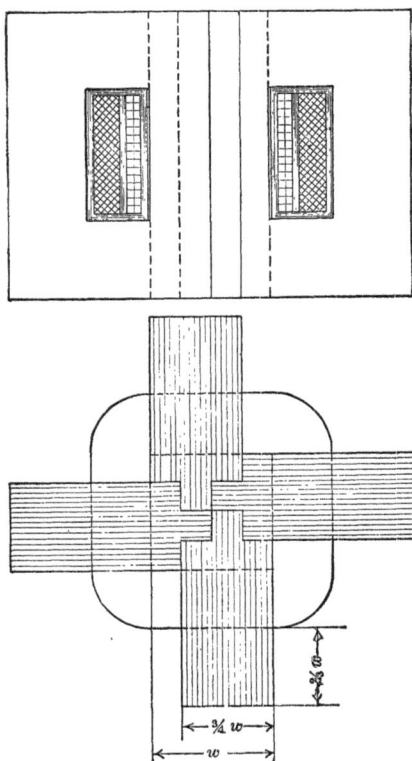

FIG. 10.—Shell-type Transformer with Distributed Magnetic Circuit.
(Square core and coil.)

again when dealing with magnetic leakage and regulation.

Fig. 12 illustrates a common arrangement of the

stampings and windings in a three-phase core-type transformer. Each of the three cores carries both primary and secondary coils of one phase. The portions

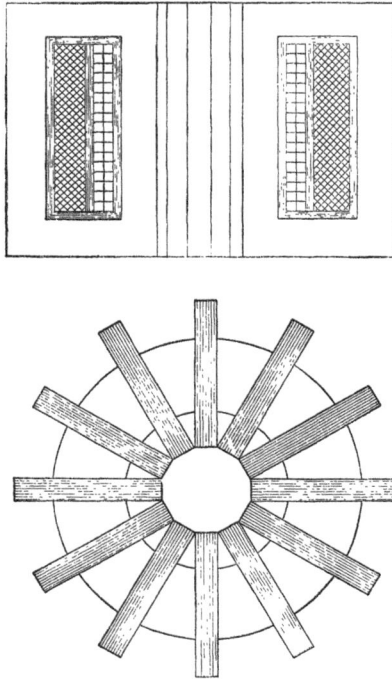

FIG. 11.—Shell-type Transformer with Distributed Magnetic Circuit. (Berry transformer with circular coil.)

of the magnetic circuit outside the coils must be of sufficient section to carry the same amount of flux as the wound cores. This will be understood if a vector diagram is drawn showing the flux relations in the various parts of the magnetic circuit. This use of certain parts of the magnetic circuit to carry the flux common to all the cores leads to a saving in material on what would be necessary for three single-phase transformers of the same total k.v.a. output; but, as men-

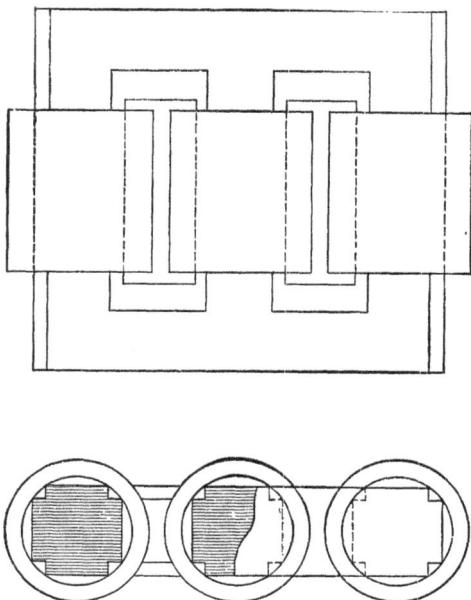

Fig. 12.—Three-phase Core-type Transformer.

tioned in Article 5, it does not follow that a three-phase transformer is always to be preferred to three separate single-phase transformers.

Figs. 13 and 14 show sections through three-phase transformers of the shell type. The former is the more common design, and it has the advantage that rectangular shaped stampings can be used throughout. The vector diagram in Fig. 13 shows how the flux Φ_c in the portion of the magnetic circuit between two sets of coils has just half the value of the flux Φ in the central core.

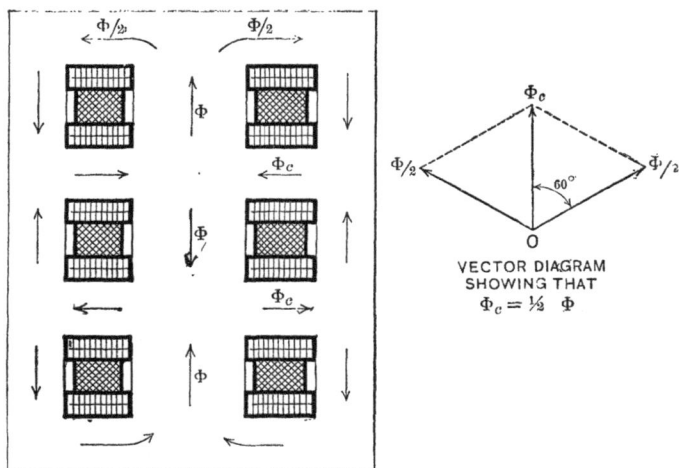

FIG. 13.—Section through Three-phase Shell Transformer. (Each phase consists of one H.T. and two L.T. coils.)

9. Mechanical Stresses in Transformers. The mechanical features of transformer design are not of sufficient importance to warrant more than a brief discussion. In the smaller transformers it is merely necessary to see that the clamps or frames securing the stampings and coils in position are sufficiently separated from the H.T. windings, and that bolts in which e.m.f.'s are likely to be generated by the main or stray magnetic fluxes are suitably insulated to prevent the establishment of electric currents with consequent I^2R losses. The tendency in all modern designs is to avoid cast iron, and use standard sections of structural steel in the assembly of the complete transformer. In this

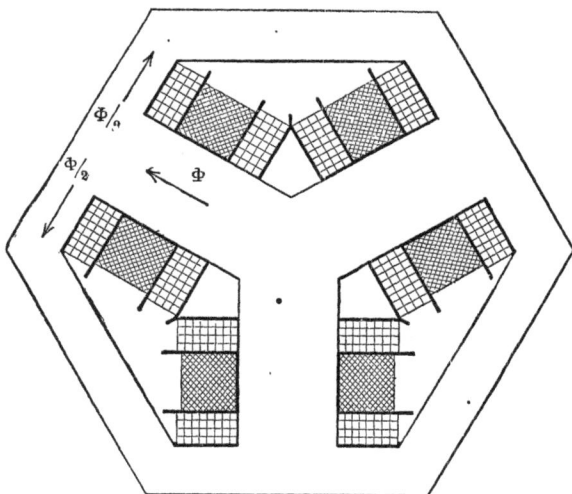

FIG. 14.—Special Design of Three-phase Shell-type Transformer.

manner the cost of special patterns is avoided and a saving in weight is usually effected. The use of standard steel sections also gives more flexibility in design, as slight modifications can be made in dimensions with very little extra cost.

In large transformers, the magnetic forces exerted under conditions of heavy overloads or· short-circuits may be sufficient to displace or bend the coils unless these are suitably braced and secured in position; and since the calculation of the stresses that have to be resisted belong properly to the subject of electrical design, it will be necessary to determine how these stresses can be approximately predetermined.

The absolute unit of current may be defined as the current in a wire which causes one centimeter length of the wire, placed at right angles to a magnetic field, to be pushed sidewise with a force of one dyne when the density of the magnetic field is one gauss.

Since the ampere is one-tenth of the absolute unit of current, we may write,

$$F = \frac{BIl}{10},$$

where F = Force in dynes;

   B = Density of the magnetic field in gausses;

   I = Current in the wire (amperes);

   l = Length of the wire (centimeters) in a direction perpendicular to the magnetic field.

It follows that the force tending to push a coil of wire of T turns bodily in a direction at right angles to a uniform magnetic field of B gausses (see Fig. 15) is

$$F = \frac{BITl}{10} \text{ dynes.}$$

If both current and magnetic field are assumed to vary periodically according to the sine law, passing through corresponding stages of their cycles at the same instant of time, we have the condition which is approximately reproduced in the practical transformer where the leakage flux passing through the windings is due to the currents in these windings.

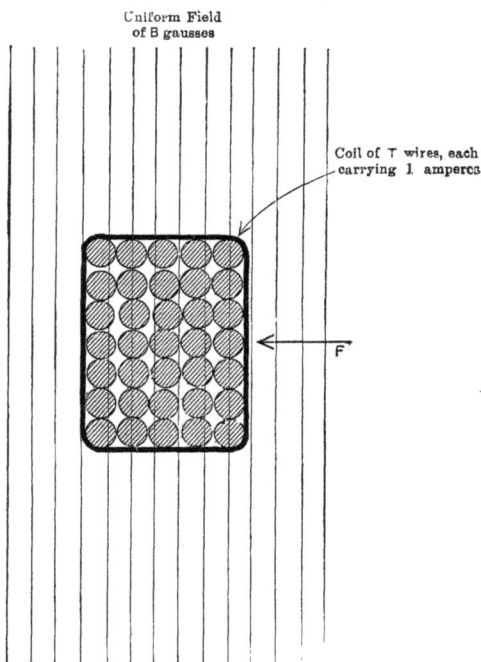

FIG. 15.—Force Acting on Coil-side in Uniform Magnetic Field.

Since the instantaneous values of the current and flux density will be $I_{\max} \sin \theta$, and $B_{\max} \sin \theta$, respectively, the average mechanical force acting upon the coil may be written,

$$F_{\text{average}} = \frac{Tl}{10} I_{\max} B_{\max} \frac{1}{\pi} \int_0^{\pi} \sin^2 \theta d\theta = \frac{Tl I_{\max} B_{\max}}{10 \times 2} \text{ dynes.}$$

If the flux density is not uniform throughout the section of coil considered, the average value of B_{max} should be taken. Let this average value of the maximum density be denoted by the symbol B_{am}. Then, since 1 lb. = 444,800 dynes, the final expression for the average force tending to displace the coil is,

$$\text{Force} = \frac{T I I_{max} B_{am}}{8,896,000} \text{ lb.} \quad \ldots \quad (4)$$

In large transformers the amount of leakage flux passing through the coils may be considerable. It will be very nearly directly proportional to I_{max}, and the mechanical forces on transformer coils are therefore approximately proportional to the square of the current. As the short-circuit current in a transformer which is not specially designed with high reactance might be thirty times the normal full-load current, the mechanical forces due to a short-circuit may be about 1000 times as great as the forces existing under normal working conditions.

Except in a few special cases, the calculation of the leakage flux is not an easy matter, and the value of B_{am} in Eq. (4) cannot usually be predetermined exactly; but it can be estimated with sufficient accuracy for the purpose of the designer, who requires merely to know approximately the magnitude of the mechanical forces which have to be resisted by proper bracing of the coils.

The calculation of leakage flux will be considered when discussing voltage regulation; but in the case of "sandwiched " coils as, for instance, in the shell type of transformer shown in Fig. 16, the distribution of the leakage flux will be generally as indicated by the diagram plotted over the coils at the bottom of the sketch.

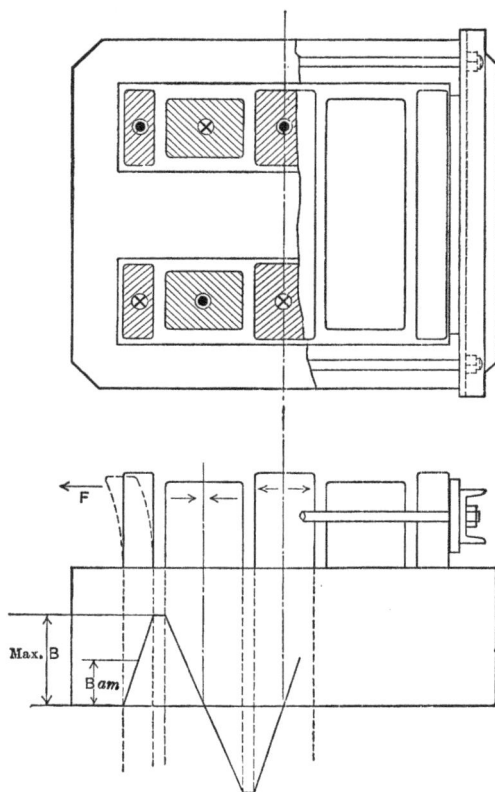

FIG. 16.—Forces in Transformer Coils Due to Leakage Flux.

When the relative directions of the currents in the primary and secondary coils are taken into account, it will be seen that all the forces tending to push the coils sidewise are balanced, except in the case of the two outside coils. In each individual coil the effect of the leakage flux is to crush the wires together; but the end

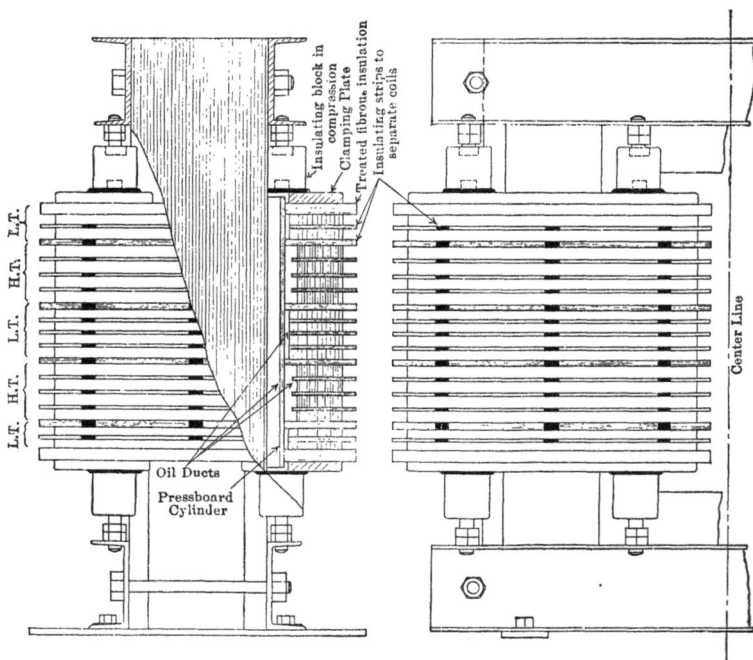

FIG. 17.—Core-type Transformer with " Sandwiched " Coils.

coils will be pushed outward unless properly secured in position.

Since there is no resultant force tending to move the windings bodily relatively to the iron stampings, a simple form of bracing consisting of insulated bars and tie rods, as shown in Fig. 16 will satisfy all requirements, and this bracing can be quite independent of the framework or clamps supporting the transformer as a whole.

In the case of core-type transformers, with rectangular coils arranged axially one within the other, the mechanical forces will tend to force the coils into a circular shape. With cylindrical concentric coils, no special bracing is necessary provided the coils are symmetrically placed axially; but if the projection of one coil beyond the other is not the same at both ends, there will be an unbalanced force tending to move one coil axially relatively to the other. If the core type of transformer is built up with flat strip " sandwiched " coils, the problem is generally similar to that of the shell type of construction. A method of securing the end coils in position with this arrangement of windings is illustrated by Fig. 17.

CHAPTER II

10. The Dielectric Circuit. Serious difficulties are not encountered in insulating machinery and apparatus for working pressures up to 10,000 or 12,000 volts, but for higher pressures (as in 150,000-volt transformers) designers must have a thorough understanding of the dielectric circuit,* if the insulation is to be correctly and economically proportioned. The information here assembled should make the fundamental principles of insulation readily understood and should enable an engineer to determine in any specific design of transformer the thicknesses of insulation required in any particular position, as between layers of windings, between high-tension and low-tension coils, and between high-tension coils and grounded metal. The data and principles outlined should also facilitate the determination of dimensions and spacings of high-tension terminals and bushings of which the detailed design is usually left to specialists in the manufacture of high-tension insulators. In presenting this information two questions are considered: (1) What is the dielectric

---

* " Insulation and Design of Electrical Windings," by A. P. M Fleming and R. Johnson—Longmans, Green & Co.
" Dielectric Phenomena in High-voltage Engineering," by F. W. Peek, Jr.—McGraw-Hill Book Company, Inc.

strength of the insulating materials used in transformer design? and (2) how can the electric stress or voltage gradient be predetermined at all points where it is liable to be excessive?

Apart from a few simple problems of insulation capable of a mathematical solution, the chief difficulty encountered in practice usually lies in determining the distribution of the dielectric flux, the concentration of which at any particular point may so increase the flux density and the corresponding electric stress that disruption of the dielectric may occur. The conception of lines of dielectric flux, and the treatment of the dielectric circuit in the manner now familiar to all engineers in connection with the magnetic circuit has made it possible to treat insulation problems * in a way that is equally simple and logical.

The analogy between the dielectric and magnetic circuits may be illustrated by Fig. 18, where a metal sphere is supposed to be placed some distance away from a flat metal plate, the intervening space being occupied by air, oil, or any insulating substance of constant specific capacity. This arrangement constitutes a condenser of which the capacity is (say) C farads. If a difference of potential of E volts is established between

---

* The dielectric circuit is well treated from this point of view in the following (among other) books:

"The Electric Circuit," by V. Karapetoff—McGraw-Hill Book Company, Inc.

"Electrical Engineering," by C. V. Christie—McGraw-Hill Book Company, Inc.

"Advanced Electricity and Magnetism," by W. S. Franklin and B. MacNutt –Macmillan Company.

the sphere and the plate, the total dielectric flux, Ψ will have to satisfy the equation

$$\Psi = EC, \quad \ldots \ldots \quad (5)$$

where Ψ is expressed in coulombs, E in volts, and C in farads.

The quantity Ψ coulombs of electricity should not be considered as a charge which has been carried from the sphere to the plate on the surface of which it remains, because the whole of the space occupied by the dielectric is actually in a state of strain, like a deflected spring, ready to give back the energy stored in it when the potential difference causing the deflection or displacement is removed. Instead, the dielectric should be considered as an electrically elastic material which will not break down or be ruptured until the " elastic limit " has been reached. The quantity Ψ, which is called the dielectric flux, may be thought of as being made up of a definite number of unit tubes of induction, the direction of which in the various portions of the dielectric field is represented by the full lines in Fig. 18. The name of the unit tube of dielectric flux is the coulomb.

If the sphere were the north pole and the plate the south pole of a magnetic circuit, the distribution of flux lines would be similar. The total flux would then be denoted by the symbol Φ, and the unit tube of induction would be called the maxwell. In place of formula (5) the following well-known equation could then be written:

$$\Phi = Mmf \times \text{permeance.} \quad \ldots \quad (6)$$

This expression is analogous to the fundamental equation for a dielectric circuit, the electrostatic capacity C being, in fact, a measure of the *permeance* of the dielectric circuit, while $\frac{1}{C}$, sometimes called the *elastance*, may be compared with reluctance in the magnetic circuit.

The dotted lines in Fig. 18 are sections through equipotential surfaces. The potential difference between

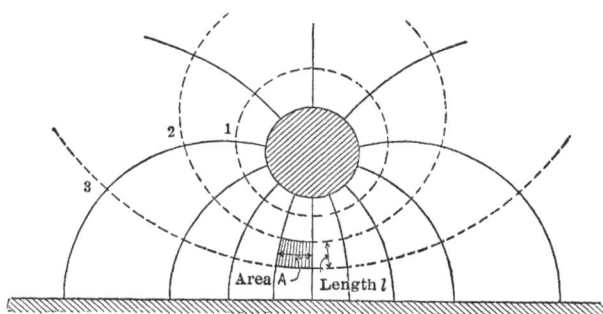

FIG. 18.—Distribution of Dielectric Flux between Sphere and Flat Plate.

any two neighboring surfaces, as drawn, is one-quarter of the total. At all points the lines of force, or unit tubes of induction, are perpendicular to the equipotential surfaces. Furthermore, the flux density, or coulombs per square centimeter, through any small portion A of an equipotential surface over which the distribution may be considered practically uniform is

$$D = \frac{\Psi}{A}. \quad \cdots \cdots \quad (7)$$

The capacity, or *permittance*,* of a small element of the dielectric circuit of length l and cross-section A is proportional to $\frac{A}{l}$, or with the proper constants inserted,

$$\text{Electrostatic capacity} = C = \left(\frac{10^9}{4\pi(3 \times 10^{10})^2}\right)\frac{kA}{l} \text{ farads} \quad (8)$$

wherein the numerical multiplier results from the choice of units. The factor k is the specific inductive capacity, or dielectric constant, of the material ($k=1$ in air), while the unit for l and A is the centimeter. This expression for capacity may conveniently be rewritten as

$$C_{mf} = \frac{8.84}{10^8}\frac{kA}{l} \text{ microfarads.} \quad . \quad . \quad . \quad (9)$$

Values of k are given in the accompanying table together with the dielectric strengths of the materials. These figures are only approximate, those referring to dielectric strength merely serving as a rough indication of what the material of average quality may be expected to withstand. The figures indicate the approximate virtual or r.m.s. value of the sinusoidal alternating voltage which, if applied between two large flat electrodes, would lead to the breakdown of a 1-cm. slab of insulating material placed between the electrodes.

What is generally understood by the *disruptive gradient*, or stress in kilovolts per centimeter, would be

---

* The reciprocal of *elastance*.

about $\sqrt{2}$ times the value given in the last column of the table. Thus, if a battery or continuous-current generator were used in the test, the pressure necessary to break down a 0.75-cm. film of air between two large flat parallel plates would be $1000 \times \sqrt{2} \times 22 \times 0.75 = 23,400$ volts.

DIELECTRIC CONSTANT AND DIELECTRIC STRENGTH OF INSULATORS

Material.	Dielectric Constant, k.	Dielectric Strength Kv. per Cm.
Air	1	22
Transformer oil	2.3	
Paper (dry)	2	80
Paper (oil impregnated)	3	100
Pressboard (dry or varnished)	3	125
Pressboard (oil impregnated)	5	150
Treated wood (across grain)	3 to 6	70
Porcelain	4.5	110
Varnished cambric	5	200
Mica	6 to 8	600
Micanite	5 to 7	300
Glass	4 to 10	90
Conductors	Infinity	

Returning to Formula (7), let *electric flux* or *quantity of electricity*, Ψ, be expressed in terms of capacity and e.m.f., with a view to determining the relation between flux density and electric stress. The Formula (8) may be written

$$C = Kk\frac{A}{l} \text{ farads,}$$

where K stands for the numerical constant. Substituting in Formula (5),

$$\Psi = E \times K k \frac{A}{l},$$

whence

$$D = K k \times \left(\frac{E}{l}\right).$$

Since $\frac{E}{l}$ is the potential gradient, or voltage drop per centimeter, which is sometimes referred to as the *electrostatic force* or *electrifying force*, and denoted by the symbol G, we may write,

$$D = K k \times G. \qquad \ldots \ldots \quad (10)$$

The analogous expression for the magnetic circuit is, $B = \mu H$.

In the case of a dielectric circuit, *electric flux density = e.m.f. per centimeter* × " *conductivity* " *of the material to dielectric flux*, while in the magnetic circuit, *magnetic flux density = m.m.f. per centimeter* × " *conductivity* " *of the material to magnetic flux*.

Since the electric stress or voltage gradient G is directly proportional (in a given material) to the flux density D, it follows that when the concentration of the flux tubes is such as to produce a certain maximum density at any point, breakdown of the insulation will occur at this point. Whether or not the rupture will extend entirely through the insulation will depend upon the value of the flux density (consequently the potential gradient) immediately beyond the limits of the local breakdown.

Given two electrical conductors of irregular shape, separated by insulating materials, the problem of calculating the capacity of the condenser so formed is very similar to that of calculating the permeance of the magnetic paths between two pieces of iron of very high permeability separated by materials of low permeability. There is no simple mathematical solution to such a problem, and the best that can be done is to fall back on the well-established law of maximum permeance, or "least resistance." According to this law the lines of force and equipotential surfaces will be so shaped and distributed that the *permittance*, or capacity, of the flux paths will be a maximum. With a little experience, ample time, and a great deal of patience, the probable field distribution can generally be mapped out, even in the case of irregularly shaped surfaces, with sufficient accuracy to emphasize the weak points of the design and to permit of the maximum voltage gradient being approximately determined.*

Before illustrating the application of the above principles in the design of transformer insulation, it will be advisable to assemble and define the quantities which are of interest to the engineer in making practical calculations.

* This method of plotting flux lines is explained, in connection with the magnetic field, at some length in the writer's book " Principles of Electrical Design."—McGraw-Hill Book Co., Inc.

Symbol:

E, e = e.m.f. or potential difference (volts);

l = length, measured along line of force (centimeters);

A = Area of equipotential surface perpendicular to lines of force (square centimeters);

$G = \dfrac{de}{dl}$ = potential gradient (volts per centimeter);

C = Capacity or permittance (farads);

$$\left(\text{farads} = \frac{\text{coulombs}}{\text{volts}} = \text{flux per unit e.m.f.}\right);$$

K = constant = 8.84×10^{-14} (farads per centimeter cube, being the specific capacity of air);

k = dielectric constant, or relative specific capacity, or permittivity ($k = 1$ for air);

Ψ = dielectric flux, or electrostatic induction ($\Psi = CE = AD$ coulombs);

D = flux density = $\dfrac{\Psi}{A}$ = KkG (coulombs per square centimeter).

11. Capacity of Plate Condenser. Imagine two parallel metal plates, as in Fig. 19, connected to the opposite terminals of a direct-current generator or battery. The area of each plate is A square centimeters and the separation between plates is l centimeters, the dielectric or material between the two surfaces being air. The edges of the plates should be rounded off to avoid concentration of flux lines. If the area A is large in comparison with the distance l, a uniform distribution of the flux Ψ may be assumed in the air gap, the density being

$$D = \frac{\Psi}{A}.$$

By Formula (9) the capacity is $C_{mf} = \dfrac{8.84 \times A}{10^8 \times l}$ micro-farads, since the specific capacity of air (k) is 1. Assuming numerical values, let $A = 1000$ sq. cm., and $l = 0.5$ cm. Then, $C = \dfrac{8.84 \times 1000}{10^{14} \times 0.5} = 1.77 \times 10^{-10}$ farads.

If $E = 10,000$ volts, the potential gradient will be $G = \dfrac{10,000}{0.5} = 20,000$ volts per centimeter. There will be

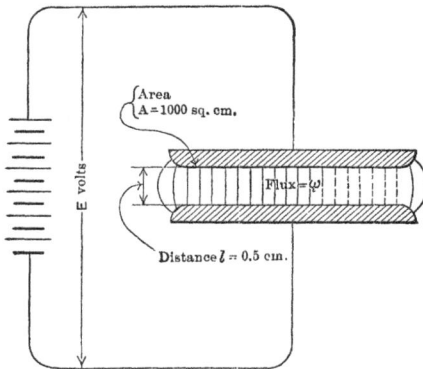

FIG. 19.—Flat Electrodes Separated by Air.

no disruptive discharge, however, because a gradient of 31,000 volts per centimeter is necessary to cause break-down in air.

By Formula (5) the total dielectric flux is $\Psi = 10,000 \times 1.77 \times 10^{-10} = 1,77 \times 10^{-6}$ coulombs.

Charging Current with Alternating Voltage. The effect of an alternating e.m.f., the crest value of which is 10,000 volts, would be to displace the above quantity of electricity $4f$ times per second, f being the frequency.

The quantity of electricity can be expressed in terms of current and time, thus, *quantity = current × time*, or *coulombs = average value of current (in amperes) during quarter period × time (in seconds) of one quarter period.* Therefore, $\Psi = I \times \left(\dfrac{2\sqrt{2}}{\pi}\right)\dfrac{1}{4f}$, where I stands for the virtual or r.m.s. value of the charging current on the sine wave assumption. Transposing terms, $I = \dfrac{4\pi f \Psi}{2\sqrt{2}}$. If E is now understood to stand for the virtual value of the alternating potential difference, $\Psi = CE \times \sqrt{2}$, whence $I = 2\pi f CE$, which is the well-known formula for calculating capacity current on the assumption of sinusoidal wave shapes.

12. Capacities in Series. When condensers are connected in parallel on the same source of voltage, the total dielectric flux is evidently determined by summing up the fluxes as calculated or measured for the individual condensers. In other words, the total capacity is the sum of the individual capacities. With condensers in series, however, the total flux, or displacement, will be the same for all the capacities in series, therefore, the calculations may be simplified just as for electric or magnetic circuits by adding the *reciprocals* of the conductance or permeance. The conception of *elastance*, corresponding to resistance in the electric circuit and reluctance in the magnetic circuit, is thus seen to have certain advantages. In the dielectric circuit

$$\text{Elastance} = \frac{1}{\text{permittance (or capacity)}} = \frac{1}{C}.$$

For a concrete example, assume that a 0.3-cm. plate of glass is inserted between the electrodes of the condenser shown in Fig. 19. The modified arrangement is illustrated by Fig. 20. On first thought it might appear that this arrangement would improve the insulation, but care must always be taken when putting layers of insulating materials of different specific inductive capacity in series, as this example will illustrate. In addition to the elastance of a 0.3-cm. layer of glass there is the

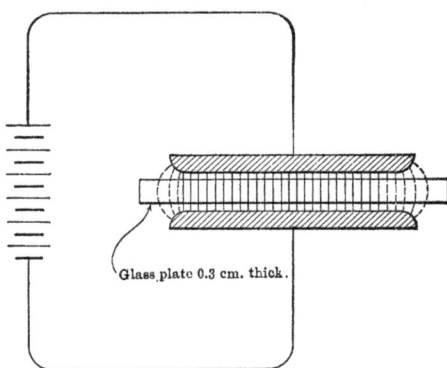

Glass plate 0.3 cm. thick.

FIG. 20.—Electrodes Separated by Air and Glass.

elastance of two layers of air of which the total thickness is 0.2 cm. Assuming that the value of the dielectric constant k for the particular quality of glass used is 7 and that G_g and G_a are the potential gradients in the glass and air respectively, then, by formula (10) $KG_a = 7KG_g$, whence $G_a = 7G_g$.

Taking the total potential difference between electrodes as 10,000 volts, the same as used in considering Fig. 19, $E = 10,000 = 0.2G_a + 0.3G_g$, whence $G_g = 5880$ volts per centimeter, and $G_a = 41,100$ volts per centimeter. Such a high gradient as $41,100$ would break down the layers of air and would manifest itself by a bluish electrical discharge between the metal plates and the glass. On the other hand, the gradient of 5880 volts per centimeter would be far below the stress necessary to rupture the glass. Nevertheless a discharge across air spaces should always be avoided in practical designs because of its injurious effect on the metal surfaces and also on certain types of insulating material. It should be observed that the introduction of the glass plate has appreciably increased the capacity of the condenser. For example, with the same voltage ($E = 10,000$) as before, the total flux is now $\Psi = AD = 1000\,(8.84 \times 10^{-14} \times 41,100) = 3.63 \times 10^{-6}$ coulombs. This value is about double the value calculated with only air between the condenser plates.

As a practical application of the principles governing the behavior of condensers in series, consider the insulation between the coils and core of an air-cooled transformer, *i.e.*, of which the coils are *not* immersed in oil. In addition assume the insulation to consist of layers of different materials made up as follows:

Material.	Total Thickness.		Dielectric Constant, k.
	Mils.	Centimeters.	
Cotton braiding and varnished cambric	70	0.178	5
Micanite........................	125	0.317	6
Pressboard...........	62	0.158	3
Air spaces (estimated).............	24	0.061	1

Then, suppose it is desired to determine how high an alternating voltage can be applied between the coils and the core before the maximum stress in the air spaces exceeds 31,000 volts per centimeter, the gradient which will cause disruption and static discharge, with the consequent danger to the insulation due to local heating and chemical action. Assuming the coil to constitute one flat plate of a condenser of which the other plate is the iron frame or core, the effect is that of a number of plate condensers in series the total elastance being

$$\frac{1}{C}=\frac{1}{C_1}+\frac{1}{C_2}+\frac{1}{C_3}+\frac{1}{C_4}.$$

By Formula (8), the individual capacities for the same surface area are proportional to $\frac{k}{l}$, and

$$\frac{KA}{C}=\frac{l_1}{k_1}+\frac{l_2}{k_2}+\frac{l_3}{k_3}+\frac{l_4}{k_4}.$$

Since

$$\frac{KA}{C}=\frac{KAE}{\Psi}=\frac{KE}{D}=\frac{KE}{KG_{air}}=\frac{E}{G_{air}},$$

the permissible maximum value of E is

$$E=31,000\left(\frac{0.178}{5}+\frac{0.317}{6}+\frac{0.158}{3}+\frac{0.061}{1}\right)$$

$$=6260 \text{ volts (maximum).}$$

The r.m.s. value of the corresponding sinusoidal alternating voltage is $\frac{6260}{\sqrt{2}}=4430$, which is the limiting potential difference between windings and grounded metal work if the formation of corona is to be avoided. A transformer having insulation made up as previously described would be suitable for a 6600-volt three-phase circuit with grounded neutral; but for higher voltages the insulation should be modified, or oil immersion should be employed to fill all air spaces. If the oil-cooled construction is employed, the previously considered insulations (slightly modified in view of possible action of the oil upon the varnish) would probably be suitable for working voltages up to 15,000.

13. Surface Leakage. A large factor of safety must be allowed when determining the distance between electrodes measured over the surface of an insulator. Whether or not spark-over will occur depends not only upon the condition of the surface (clean or dirty, dry or damp), but also upon the shape and position of the terminals or conductors. It is therefore almost impossible to determine, other than by actual test, what will happen in the case of any departure from standard practice. Surface leakage occurs under oil as well as in air, but generally speaking, the creepage distance under oil need be only about one-quarter of what is necessary in air.

An important point to consider in connection with surface leakage is illustrated by Figs. 21 and 22. In Fig. 21, a thin disk of porcelain (or other solid insulator) separates the two electrodes, while in Fig. 22, the same material is in the form of a thick block providing a leakage path (l) of exactly the same length as in Fig. 21. The voltage required to cause spark-over will be con-

siderably greater for the block of Fig. 22 than for the
disk of Fig. 21. This condition exists because the flux
concentration due to the nearness of the terminals in
Fig. 21 begins breaking down the layers of air around
the edges of the electrodes at a much lower total potential difference than will be necessary in the case of the
thicker block of Fig. 22. The effect of the incipient
breakdown is, virtually, to make a conductor of the air

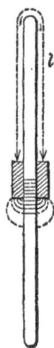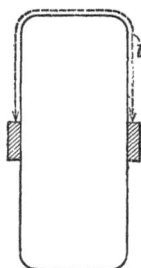

FIG. 21.   FIG. 22.

FIG. 21.—Surface Leakage over Thin Plate.
FIG. 22.—Surface Leakage Over Thick Insulating Block.

around the edges of the metal electrodes, and a very
slight increase in the pressure will often suffice to break
down further layers of air and so result in a discharge
over the edges of the insulating disk. The phenomenon
of so-called surface leakage may thus be considered
as largely one of flux concentration or potential gradient. Sometimes it will be easier to eliminate trouble
due to surface leakage by altering the design of ter-

minals and increasing the *thickness* of the insulation than by adding to the length of the creepage paths.

14. Practical Rules Applicable to the Insulation of High-voltage Windings. For working pressures up to 16,000 volts, solid insulation, including cotton tape, micanite, pressboard, horn paper, or any insulating material of good quality used to separate the windings from the core or framework, should have a total thickness of approximately the following values:

Voltage.	Thickness of Insulation (Mils)
110	40
400	45
1,000	65
2,200	90
6,600	180
12,000	270
16,000	350

In large high-voltage power transformers, cooled by air blast, the air spaces are relied upon for insulation. The clearances between coils and core or case are necessarily much larger than in oil-cooled transformers, and calculations similar to the example previously worked out should be made to determine whether or not the insulation is sufficient and suitably proportioned to prevent brush discharge. The calculations are made on the basis of several plate condensers in series; thus the flux density and dielectric stress in the various layers of insulation can be approximately predetermined. The difficulty of avoiding static discharges will generally stand in the way of designing economical air-cooled transformers for pressures much in excess of 30,000 volts. A rough rule for air clearance is to allow a distance equal to $\frac{kv+1}{4}$ inches, where kv stands for the virtual value of the alternating potential difference in kilovolts between the two surfaces considered.

With oil-immersed transformers, the oil channels should be at least 0.25 in. wide in order that there may be free circulation of the oil. In high-voltage trans- formers having a considerable thickness of insulation between coils and core, it is advantageous to divide the oil spaces by partitions of pressboard or similar material. Assuming the total thickness of oil to be no greater than that of the solid insulation, a safe rule is to allow 1 mil for every 25 volts. For instance, a total thickness of insulation of 1 in. made up of 0.5 in. of solid insulation and two 0.25 in. oil ducts would be suitable for a working pressure not exceeding $25 \times 1000 = 25,000$ volts. Further particulars relating to oil insulation will be given later.

It is customary to limit the volts per coil to 5000, and the volts between layers of winding to 400. Special attention must be paid to the insulation under the finishing ends of the layers by providing extra insulation ranging from thin paper to Empire cloth or even thin fullerboard, the material depending upon the voltage and also upon the amount of mechanical protection required to prevent cutting through the insulation where the wires cross. Sometimes the insulation is bent around the end wires of a layer to prevent breakdown over the ends of the coil.   Where space permits, however, the layers of insulation may be carried beyond the ends of the winding so as to avoid surface leakage.   This arrangement is more easily carried out in core-type transformers than in shell-type units.   A practical rule for determining the surface distance (in inches) required to prevent leakage (given by Messrs. Fleming and Johnson in the book previously referred to) is to allow 0.5 in. $+0.5\times$ kilovolts, when the surfaces are in air.   For surfaces under oil, the allowance may be $0.5+0.1\times$ kilovolts. In any case it is important to see that the creepage surfaces are protected as far as possible from deposits of dirt.   When the coils of shell-type transformer are "sandwiched," it is customary to use half the normal number of turns in the low-tension coils at each end of the stack.   This has the advantage of keeping the high-tension coils well away from the iron stampings and clamping plates or frame.

Extra Insulation on End Turns.   Concentration of potential between turns at the ends of the high-tension winding is liable to occur with any sudden change of voltage across the transformer terminals, such as when the supply is switched on, or when lightning causes potential disturbances on the transmission lines.   It is, therefore, customary to pay special attention to the insulation of the end turns of the high-tension winding. Transformers for use on high-voltage circuits usually have about 75 ft. at each end of the high-tension winding insulated to withstand three to four times the voltage between turns that would puncture the insulation in the body of the winding.

It is very difficult to predetermine the extra pressure to which the end turns of a power transformer connected to an overhead transmission line may at times be subjected, but it is safe to say that the instantaneous potential difference between turns may occasionally be of the order of forty to fifty times the normal working pressure. In such cases the usual strengthening of the insulation on the end turns would not afford adequate protection, and for this reason a separate specially designed reactance coil connected to each end of the high-tension winding would seem to be the best means of guarding against the effects of surges or sudden changes of pressure occurring in the electric circuit outside the transformer. The theory of abnormal pressure rises in the end sections of transformer windings will not be discussed here.

15. Winding Space Factor. Knowing the thickness of the cotton covering on the wires, the insulation between layers of winding, between coil and coil and between coil and iron stampings, it becomes an easy matter to determine approximately the total cross-section of the winding-space to accommodate a given cross-section of copper. The ratio $\dfrac{\text{cross-section of copper}}{\text{cross-section of winding space}}$, which is known as the space factor, will naturally decrease with the higher voltages and smaller sizes of wire. This factor may be as high as 0.46 in large transformers for pressures not exceeding 2200 volts; in 33,000-volt transformers for outputs of 200 k.v.a. and upward it will have a value ranging between 0.35 and 0.2, while in oil-immersed power transformers for use on 100,000-volt circuits the factor may be as low as 0.06.

16. Oil Insulation. There is a considerable amount of published matter relating to the properties of insulating oils, and also to the various methods of testing, purifying, and drying oils for use in transformers. A concise statement of the points interesting to those installing or having charge of transformers will be found in W. T. Taylor's book on transformers.* What follows here is intended merely as a guide to the designer in providing the necessary clearances to avoid spark-over, including a reasonable factor of safety.

Mineral oil is generally employed for insulating purposes, its main function in transformers being to transfer the heat by convection from the hot surfaces to the outside walls of the containing case, or to the cooling coils when these are provided. The presence of an extremely small percentage of water reduces the insulating properties of oil considerably. It is therefore important to test transformer oil before using it, and if necessary extract the moisture by filtering through dry blotting paper, or by any other approved method. Dry oil will withstand pressures up to 50,000 volts (alternating) between brass disks 0.5 in. in diameter with a separation of 0.2 in. For use in high-voltage transformers, the oil should be required to withstand a test

---

* "Transformer Practice," by W. T. Taylor—McGraw-Hill Book Company, Inc. For further information refer H. W. Tobey on the "Dielectric Strength of Oil"—*Trans.* A.I.E.E.; Vol. **XXIX**, page 1189 (1910). Also "Insulating Oils," *Journ. Inst. E.E.,* Vol. 54, page 497 (1916).

of 45,000 volts under the above conditions. The good insulating qualities of oil suggest that only small clearances would be required in transformers, even for high voltages; but the form of the surfaces separated by the layer of oil will have a considerable effect upon the concentration of flux density, and therefore upon the voltage gradient. As an example, if 100,000 volts breaks down a 1-in. layer of a certain oil between two parallel disks 4 in. in diameter, the same pressure will spark across a distance of about 3.5 in. between a disk and a needle point.

Partitions of solid insulation such as pressboard or fullerboard are always advisable in the spaces occupied by the oil, since they will prevent the lining up of partly conducting impurities along the lines of force and reduce the total clearance which would otherwise be necessary.

In a transformer oil of average quality, the sparking distance between a needle point and a flat plate is approximately $(0.25+0.04\times\text{kv.})$ inches. Since there may be sharp corners or irregularities corresponding to a needle point, which will produce concentration of dielectric flux, it therefore seems advisable to introduce a factor of safety for oil spaces between high tension and grounded metal—for instance, between the ends of high-tension coils and the containing case—by basing the oil space dimension on the formula,

$$\text{Thickness of oil (inches)} = 0.25 + 0.08 \times \text{kv.,} \qquad (11)$$

where kv. stands for the working pressure in kilovolts.

With two or three partitions of solid insulating material dividing the oil space into sections, the total thickness need not exceed

$$0.25 + 0.05 \times kv. \quad . \quad . \quad . \quad . \quad (12)$$

If the total thickness of solid insulation is about equal to that of the oil ducts (not an unusual arrangement between coils and core), the rule previously given for solid insulation may be slightly modified to include a minimum thickness of 0.25 in., and put in the form,

Total thickness of oil ducts plus ⎫
solid insulation of approxi- ⎬ $= 0.25 + 0.03 \times kv.$ (13)
mately equal thickness (inches) ⎭

A suitable allowance for surface leakage under oil, in inches, as already given, is

$$0.5 + 0.1 \times kv. \quad . \quad . \quad . \quad . \quad (14)$$

17. Terminals and Bushings. The exact pressure which will cause the breakdown of a transformer terminal bushing generally has to be determined by test, because the shape and proportions of the metal parts are rarely such that the concentration of flux density at corners or edges can be accurately predetermined.*

---

* The reader who desires to go deeply into the study of high-pressure terminal design should refer to the paper by Mr. Chester W. Price entitled " An Experimental Method of Obtaining the Solution of Electrostatic Problems, with Notes on High-voltage Bushing Design." *Trans.* A.I.E.E., Vol. 36, page 905 (Nov., 1917).

However, there are certain important points to bear in mind when designing the insulation of transformer terminals, and these will now be referred to briefly.

The high-tension leads of a transformer may break down (1) by puncture of the insulation, or (2) by spark-over from terminal to case. If the transformer lead could be considered as an insulated cable with a suitable dielectric separating it from an outer concentric

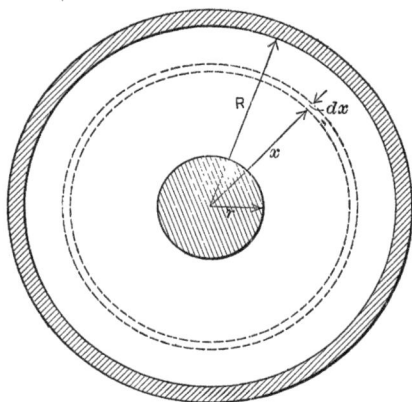

FIG. 23.—Section through Insulated Conductor.

metal tube of considerable length, the calculation of the puncture voltage (1) would be a simple matter. For instance, let r in Fig. 23 be the radius of the inner (cylindrical) conductor, and R the internal radius of the enclosing tube, the space between being filled with a dielectric of which the specific inductive capacity (k) is constant throughout the insulating material. The equipotential surfaces will be cylinders, and the flux density over the surface of any cylinder of radius x and of length 1 cm., will be $D = \dfrac{\Psi}{2\pi x}$.

By Formula (10) the potential gradient is,

$$G = \frac{D}{Kk} = \frac{\Psi}{2\pi x Kk}. \quad \cdots \quad (15)$$

In order to express this relation in terms of the total voltage E, it is necessary to substitute for the symbol Ψ its equivalent $E \times C$, and calculate the capacity C of the condenser formed by the rod and the concentric tube. Considering a number of concentric shells in series, the elastance may be written as follows:

$$\frac{1}{C} = \int_r^R \frac{dx}{2\pi x Kk} = \frac{1}{2\pi Kk} \log_e \frac{R}{r}. \quad \cdots \quad (16)$$

Substituting in (15), we have,

$$G = \frac{E}{x \log_e \dfrac{R}{r}} \text{ volts per centimeter}, \quad \cdots \quad (17)$$

the maximum value of which is at the surface of the inner conductor, where

$$G_{max} = \frac{E}{r \log_e \dfrac{R}{r}}. \quad \cdots \quad (18)$$

This formula is of some value in determining the thickness of insulation necessary to avoid overstressing the

dielectric; but it is not strictly applicable to transformer bushings in which the outer metal surface (the
bushing in the lid of the containing tank) is short in
comparison with the diameter of the opening. The
advantage of having a fairly large value for r is indicated
by Formula (18), and a good arrangement is to use a
hollow tube for the high-tension terminal, with the lead
from the windings passing up through it to a clamping
terminal at the top.

Solid porcelain bushings with cither smooth or corrugated surfaces may be used for any pressure up to
40,000 volts, but for higher pressures the oil-filled type
or the " condenser " type of terminal is preferable. In
designing plain porcelain bushings it is important to
see that the potential gradient in the air space between
the metal rod and the insulator is not liable to cause
brush discharge, as this would lead to chemical action,
and a green deposit of copper nitrate upon the rod. The
calculations would be made as explained for the parallel-
plate condensers in which a sheet of glass was inserted
(see " Capacities in Series "), except that the elastances
of the condensers are now expressed by Formula (16).

18. Oil-filled Bushings. The chief advantages of a
hollow insulating shell filled with oil or insulating compound that can be poured in the liquid state, are the
absence of air spaces where corona may occur, and the
possibility of obtaining a more uniform and reliable
insulation than with solid insulators—such as porcelain,
when the thickness is considerable. The metal ring
by which such an insulator (see Fig. 24) is secured to the
transformer cover usually takes the form of a cylinder

of sufficient length to terminate below the surface of the
oil. The advantage of this arrangement is that the
dielectric flux over the surface of the lower part of the
insulator is through oil only, and not as would otherwise
be the case, through oil and air. With the two materials of different dielectric constants, the stress at the
surface of the oil may exceed the dielectric strength of
air, in which case there would be corona or brush discharge which might practically short-circuit the air
path and increase the stress over that portion of the
surface which is under the oil.

The bushing illustrated in Fig. 24 has been designed
for a working pressure of 88,000 volts between high-
tension terminal and case, the method of computation
being, briefly, as follows: Applying the rule for surface leakage distances previously given, this dimension
is found to be $0.5 + \frac{88}{2} = 44.5$ in. The insulator need
not, however, measure 44.5 in. in height above the
cover of the transformer case, because corrugations can
be used to obtain the required length. A safe rule to
follow in deciding upon a minimum height, *i.e.*, the
direct distance in air between the terminal and the
grounded metal, is to make this dimension at least as
great as the distance between needle points that would
just withstand the test voltage without sparking over.
The test pressure is usually twice the working pressure
plus 1000 volts, or 177 kv. (r.m.s. value) in this particular case. This value corresponds to a distance of
about 48 cm., or (say) 19 in. In order that there may
be an ample margin of safety, it will be advisable to
make the total height of the insulator not less than 22

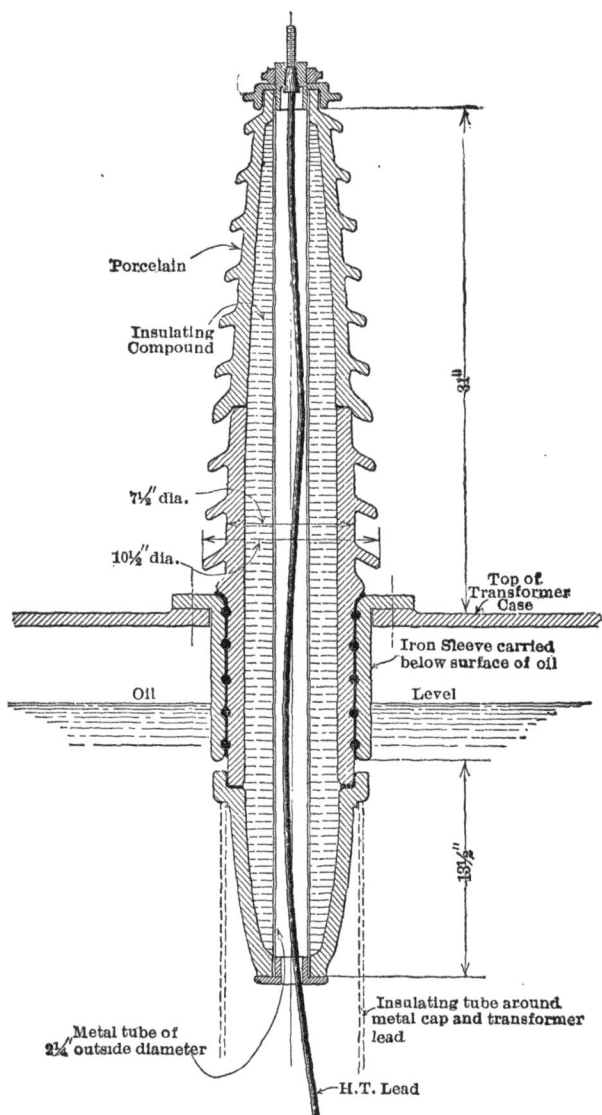

Porcelain

Insulating
Compound

$7\frac{1}{2}''$ dia.

$10\frac{1}{2}''$ dia.

$31\frac{1}{2}''$

Top of
Transformer
Case

Iron Sleeve carried
below surface of oil

Oil   Level

$13\frac{1}{2}''$

Metal tube of
$2\frac{1}{4}''$ outside diameter

Insulating tube around
metal cap and transformer
lead

H.T. Lead

FIG. 24.—Three-part Composition-filled Porcelain Transformer Bushing,
Suitable for a Working Pressure of 88,000 Volts to Ground.

in., apart from the number or depth of the corrugations. The actual height in Fig. 24 is 31 in. because the corrugations on the outside of the porcelain shell are neither very numerous nor very deep. In this connection it may be stated that a short insulator with deep corrugations designed to provide ample surface distance is not usually so effective as a tall insulator with either a smooth surface or shallow corrugations. The reason is that much of the dielectric flux from the high-tension terminal to the external sleeve or supporting framework passes through the flanges, the specific inductive capacity of which is two to three times that of the air between them. The result is an increased stress in the air spaces, which is equivalent to a reduction in the effective height of the insulator.

In the design under consideration it is assumed that the hollow (porcelain) shell is filled with an insulating compound which is solid at normal temperatures, and that the joints therefore need not be so carefully made as when oil is used. The insulator consists of three parts only, which are jointed as indicated on the sketch. Oil-filled bushings for indoor use generally have a large number of parts, usually in the form of flanged rings with molded tongue-and-groove joints filled with a suitable cement. There is always the danger, however, that a vessel so constructed may not be quite oil-tight, therefore the solid compound has an advantage over the oil in this respect.

The creepage distance over the surface of the insulator in oil may be very much less than in air. Applying the rule previously given, the minimum distance in this case would be $0.5 + (0.1 \times 88) = 9.3$ in. In the design illustrated by Fig. 24, however, this dimension has been increased about 50 per cent with a view to keeping the high-tension connections well away from the surface of the oil and grounded metal. To prevent the accumulation of conducting particles in the oil along the lines of stress, and afford increased protection with only a small addition in cost, it is advisable to slip one or more insulating tubes over the lower part of the terminal, as indicated by the dotted lines in the sketch. Corrugations on the surface of the insulator in the oil are usually unnecessary, and sometimes objectionable because they collect dirt which may reduce the effective creepage distance.

Having decided upon the height and surface distances to avoid all danger of spark-over, the problem which remains to be dealt with is the provision of a proper thickness of insulation to prevent puncture. In order to avoid complication of the problem by considering the different dielectric constants (k) of the compound used for filling and of the external shell (assumed in this case to be porcelain), it may be assumed either that there is no difference in the dielectric constants of the two materials, or that the thickness of the inclosing shell of porcelain is negligibly small in relation to the total external diameter of the insulator. Either assumption, neglecting the error due to the limited length of the external metal sleeve,* permits the use of Formula (18),

* The maximum stress in the dielectric might be 5 to 10 per cent greater than calculated by using formulas relating to very long cylinders. The corners at the ends of the outer cylinder should be rounded off to avoid concentration of dielectric flux at these places.

giving the relation between the maximum potential gradient and the dimensions of the bushing, without correction.

Suppose that the disruptive gradient of the insulating compound is 90 kv. per centimeter (maximum value) or 63.5 kv. per centimeter (r.m.s. value) of the alternating voltage. With a test pressure of 177 kv. and a margin of safety of 25 per cent, the value of E in Formula (18) will therefore be $E = 177 \times 1.25 \times \sqrt{2} = 313$ kv.

Since the disadvantage of a very small value of r is evident from an inspection of the formula, the outside diameter of the inner tube is made 2.25 in. Then, since

$$G = \frac{E}{r \log_e \dfrac{R}{r}}$$

$$\log_{10} \frac{R}{r} = \frac{313}{2.54 \times 1.125 \times 90 \times 2.303} = 1.216,$$

whence $R = 3.79$, or (say) 3.75 in. An external diameter of 7.5 in. at the center of the insulator will therefore be sufficient to prevent the stress at any point exceeding the rupturing value even under the test pressure.

19. Condenser Type of Bushing. If the total thickness of the insulation between the high-tension rod and the (grounded) supporting sleeve is divided into a number of concentric layers by metallic cylinders, the concentration of dielectric flux at certain points (leading to high values of the voltage gradient) is avoided. The bushing then consists of a number of plate condensers in series, with a definite potential difference between the plates. If the total radial depth of insulation is divided into a large number of concentric layers (of the same thickness), separated by cylinders of tinfoil (of the same area), the several condensers would all have the same capacity. The dielectric flux density, and therefore the potential gradient, would then be the same in all the condensers, so that the outer layers of insulation would be stressed to the same extent as the inner layers, and the total radial depth of insulation would be less than when the stress distribution follows the logarithmic law (Formula 18) as in the case of the solid porcelain, or oil-filled, bushing.

The section on the right-hand side of Fig. 25 is a diagrammatic representation of a condenser bushing shaped to comply with the assumed conditions of equal thicknesses of insulation and equal areas of the condenser plates. With a sufficient number of concentric layers, the condition of equal potential difference between plate and plate throughout the entire thickness would be approximated; but the creepage distance over the insulation between the edges of the metal cylinders would be much smaller for the outer layers than for layers nearer to the central rod or tube. It is equally, if not more, important to prevent excessive stress over the surface than in the body of the insulator, and a practical condenser type of terminal can be designed as a compromise between the two conflicting requirements. By making the terminal conical in form, as indicated by the dotted lines on the right-hand side and the full lines on the left-hand side of the sketch (Fig. 25), neither of the ideal conditions will be exactly

fulfilled, but practical terminals so constructed are easily
manufactured, and give satisfaction on circuits up to

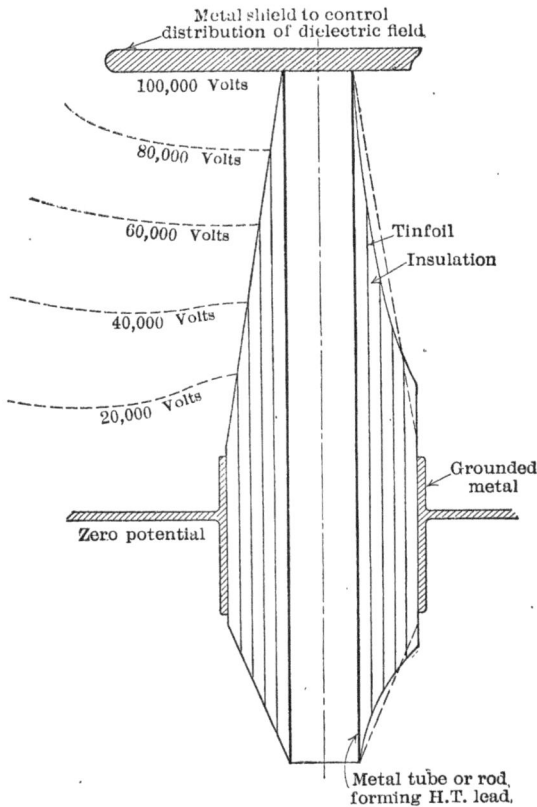

FIG. 25.—Illustrating Principle of Condenser Type Bushing.

150,000 volts. By varying the thickness of the individual insulating cylinders, it is an easy matter to
design a condenser type terminal of which the con-

densers in series all have the same capacity even while
the outside surface is conical in shape as shown on the
left-hand side of Fig. 25.   This gives a uniform potential
gradient along the surface, and results in a good practical
form of condenser-type bushing.

If the ends of the metal cylinders coincide with equipotential surfaces having the same potential as that
which they themselves attain by virtue of the respective
capacities of the condensers in series, there will be no
corona or brush discharge at the edges of these cylinders.   This ideal condition is represented diagrammatically in Fig. 25, where a large metal disk is shown
at the top of the terminal.   The object of this metal
shield is to distribute the field between the terminal
and the transformer cover in such a manner as to satisfy
the above-mentioned condition.   In practice, the tendency for corona to form at the exposed ends of the tinfoil cylinders is counteracted by treating the finished
terminal with several coats of varnish, and surrounding
it with an insulating cylinder filled with an insulating
compound which can be poured in the liquid form and
which solidifies at ordinary temperatures.   This construction is shown in Fig. 26, which represents a practical terminal of the condenser type.   Compared with
Fig. 24, it is longer, but appreciably smaller in diameter
where it passes through the transformer cover.

The dimensions of a condenser-type terminal such
as illustrated in Fig. 26 may be determined approximately as follows: Assuming the working pressure as
88,000 volts, and the maximum permissible potential
gradient in the dielectric (usually consisting of tightly·

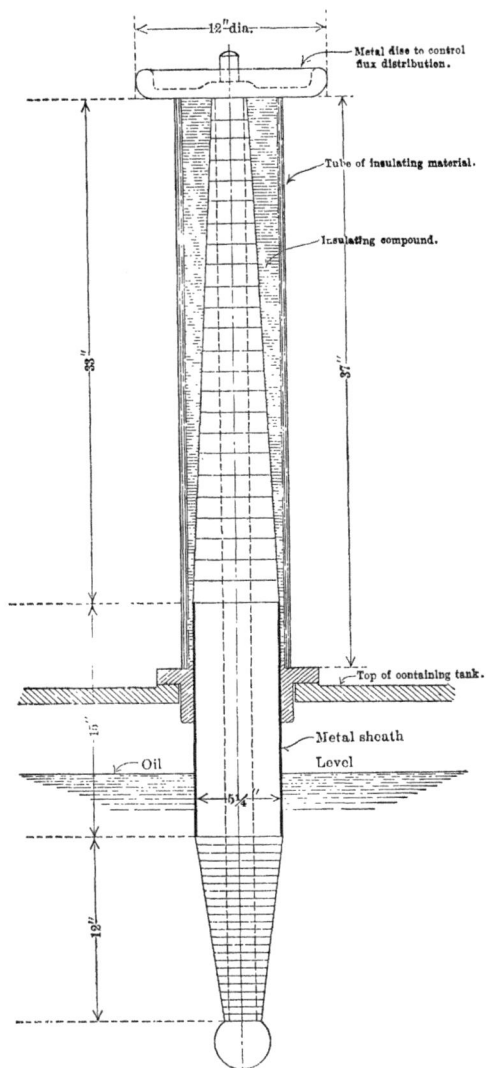

FIG. 26.—Condenser-type Transformer Bushing Suitable for a Working Pressure of 88,000 Volts.

wound layers of specially treated paper) as 90 kv.,* the maximum radial thickness of insulation required will be $\dfrac{\text{total volts}}{\text{voltage gradient}} = \dfrac{313}{90} = 3.48$ cm. or (say) 1.5 in. to include an ample allowance for the dividing layers of metal foil. If the inner tube is 2.25 in. in diameter, as in the previous example, the external diameter over the insulation at the center will be 2.25 $\times 3 = 5.25$ in. instead of the 7.5 in. required for the previous design.

It is customary to allow about 4000 volts per layer, and twenty-two layers of insulation alternating with twenty-two layers of tinfoil are used in this particular design. It is true that ideal conditions will not be actually fulfilled; the aggregate thickness of insulation might have to be slightly greater than 1.5 in., but the inner tube might be made 1.75 in. or 2 in. instead of 2.25 in., and a practical terminal for 88,000-volt service could undoubtedly be constructed with a diameter over the insulation not exceeding 5.25 in.

The projection of the terminal above the grounded plate (the cover of the transformer case) need not be so great as would be indicated by the application of the practical rule previously given for surface leakage distance, namely, that this distance should be $\left(0.5 + \dfrac{\text{kv.}}{2}\right)$ in., where kv. stands for the working pressure. The reason why a somewhat shorter distance is permissible is that the surface of the terminal proper has been covered by varnish and a solid compound, and so far as the enclosing cylinder is concerned, the stress along the sur-

* Same as in the example of the compound-filled insulator.

face of this cylinder will be fairly uniform, especially if a large flux-control shield is provided, as shown in Fig. 26. In order to avoid the formation of corona at the lower terminal (below the surface of the oil) this end may conveniently be in the form of a sphere, the diameter of which would depend upon the voltage and the proximity of grounded metal.

The following particulars relate to a condenser type bushing actually in service on 80,000 volts. The layers of insulation are built up on a metal tube of 2.25 in. outside diameter. The diameter over the outside insulating cylinder is 5.3 in. This bushing has uniform capacity, the thickness of the inner and outer insulating wall being the same, namely 0.062 in.; but the thickness of the intermediate cylinders is variable, the maximum being 0.073 in. for the twelfth and thirteenth cylinders. (A plot of the individual thickness forms a hyperbolic curve.) The static shield or " hat " is 9 in. diameter and 2 in. thick, the edge being rolled to a true semicircle. When provided with a casing filled with gum, and when the taper is such that the steps on the air end are 1.69 in. (total length = $1.69 \times 22 = 37.2$ in.), there is no difficulty in raising the voltage to 300,000 (r.m.s. value) without arc-over. The same bushing without a casing would arc-over at about 285,000 volts; but this can be raised to the same value as for the terminal with gum-filled casing if the size of the static shield is increased to about 2 ft. diameter. When the arc-over voltage is reached, the discharge takes place between the edge of the static shield and the flange which is bolted to the transformer case.

CHAPTER III

· **20. Losses in Core and Windings.** The power loss in the iron of the magnetic circuit is due partly to hysteresis and partly to eddy currents. The loss due to hysteresis is given approximately by the formula

$$\text{Watts per pound} = K_h B^{1\cdot6} f,$$

where K_h is the hysteresis constant which depends upon the magnetic qualities of the iron. The symbols B and f stand, respectively, for the maximum value of the magnetic flux density, and the frequency. An approximate expression for the loss due to eddy currents is

$$\text{Watts per pound} = K_e \ (Bft)^2,$$

where t is the thickness of the laminations, and K_e is a constant which is proportional to the electric conductivity of the iron.

With the aid of such formulas, the hysteresis and eddy current losses may be calculated separately, and then added together to give the total watts lost per pound of the core material; but it is more convenient to use curves such as those of Fig. 27, which should be plotted

69

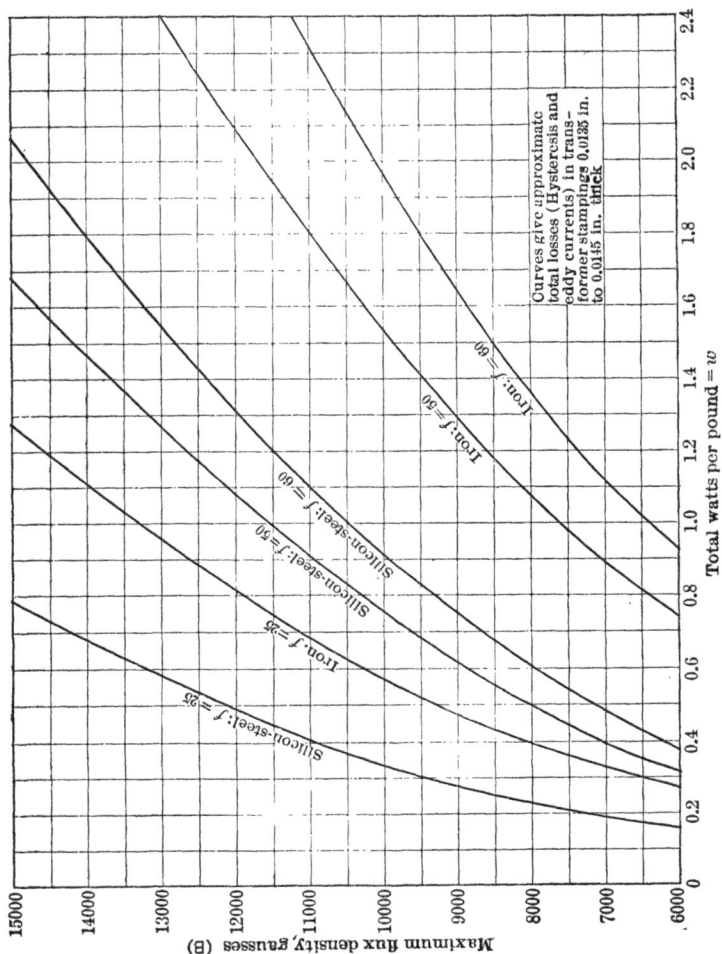

FIG. 27.—Losses in Transformer Stampings.

from tests made on samples of the iron used in the construction of the transformer. These curves give the relation between maximum value of flux density, and total iron loss per pound at various frequencies. The curves of Fig. 27 are based on average values obtained with good samples of commercial transformer iron and silicon-steel; the thickness of the laminations being about 0.014 in.

The cost of silicon-steel stampings is greater than that of ordinary transformer iron; but the smaller total iron loss resulting from the use of the former material will almost invariably lead to its adoption on economic grounds. The eddy-current losses are smaller in the alloyed material than in iron laminations of the same thickness because of the higher electrical resistance of the former. • The permeability of silicon-steel is slightly lower than that of ordinary iron, and this may lead to a somewhat larger magnetizing current; on the other hand, the modern alloyed transformer material (silicon-steel) is non-ageing, that is to say, it has not the disadvantage common to transformers constructed fifteen to twenty years ago, in which the iron losses increased appreciably during the first two or three years of operation. The " ageing " of the ordinary brands of transformer iron— resulting in larger losses—is caused by the material being maintained at a fairly high temperature for a considerable length of time.

The maximum flux density in transformer cores is generally kept below the knee of the B-H curve. As a guide for use in preliminary designs, usual values of B (gausses) are given below:

APPROXIMATE VALUES OF B IN TRANSFORMER CORES

	$f = 25$	$f = 50$ or 60
Small lighting or distributing transformers:		
Ordinary iron............	8,000 to 11,000	5,000 to 7,000
Alloyed iron.............	10,000 to 13,000	8,000 to 11,000
Power transformers:		
Ordinary iron.............	10,000 to 13,000	9,000 to 11,000
Alloyed iron.............	11,000 to 14,000	11,000 to 14,000

The losses in the iron core are usually less than one watt per pound, although they sometimes amount to 1.5 watts, and even 1.8 watts, per pound. The higher figures apply to large, artificially cooled, power transformers.

Current Density in Windings. Even with well-ventilated coils (air blast), or improved methods of producing good oil circulation, the permissible current density in the copper windings is limited by local heating. If the watts lost per pound of copper exceed a certain amount, there will be danger of internal temperatures sufficiently high to cause injury to the insulation. As a rough guide in deciding upon suitable values for trial dimensions in a preliminary design, the following approximate figures may be used:

AVERAGE VALUES OF CURRENT DENSITY (Δ) IN
COMMERCIAL TRANSFORMERS

Type of Transformer.	Amperes per Square Inch.
Standard lighting transformers (oil-immersed; self-cooled).	800 to 1200
Transformers for use in Central Generating Stations, or Substations (oil-cooled, or air blast).................	1100 to 1600
Large, carefully designed transformers, oil-insulated, with forced circulation of oil, or with water cooling-coils.....	1400 to 1900

When the current is very large, it is important to sub-divide the conductors to prevent excessive loss by eddy currents. When flat strips are used, the laminations must be in the direction of the leakage flux lines. It is advisable to add from 10 to 15 per cent to the calculated I^2R loss when the currents to be carried are large, even after reasonable precautions have been taken to avoid large local currents by subdividing the conductors.

The mere subdivision of a conductor of large cross-section does not always eliminate the injurious effects of local currents in the copper, because, unless each of the several conductors that are joined in parallel at the terminals does not enclose the same amount of leakage flux, there will be different e.m.f.'s developed in various sections of the subdivided conductor, and consequent lack of uniformity in the current distribution. This objection can sometimes be overcome by giving the assembled conductor (of many parallel wires or strips) a half twist, and so changing the position of the individual conductors relatively to the leakage flux; but, in any case, once this cause of increased copper loss is recognized, it is generally possible to dispose and join together the several elements of a compound conductor so that the leakage flux shall affect them all equally.

21. Efficiency. The output of a single-phase transformer, in watts, is

$$W = E_s I_s \cos \theta,$$

where E_s is the secondary terminal voltage; I_s, the secondary current; and $\cos \theta$, the power factor of the secondary load. The *percentage efficiency* is then:

$$100 \times \frac{W}{W + \text{iron losses} + \text{copper losses}}.$$

All-day Efficiency. The all-day efficiency is a matter of importance in connection with distributing transformers, because, although the amount of the copper loss falls off rapidly as the load decreases, the iron loss continues usually during the twenty-four hours, and may be excessive in relation to the output when the transformer is lightly loaded, or without any secondary load, during many hours in the day.

What is understood by the *all-day percentage efficiency* is the ratio given below, the various items being calculated or estimated for a period of twenty-four hours:

$$\frac{100 \times \text{Secondary output in watt-hours}}{\text{Sec. watt-hrs.} + \text{watt-hrs. iron loss} + \text{watt-hrs. copper loss}}.$$

It is in order that this quantity may be reasonably large that the iron losses in distributing transformers are usually less than in power transformers designed for the same maximum output.

Efficiency of Modern Transformers. The alternating-current transformer is a very efficient piece of apparatus, as shown by the following figures which are an indication of what may be expected of well-designed transformers at the present time.

FULL-LOAD EFFICIENCIES OF SMALL LIGHTING TRANS-
FORMERS FOR USE ON CIRCUITS UP TO 2200 VOLTS

Output, k.v.a.	Efficiency (per cent)
1	From 94.1 to 96
2	From 94.6 to 96.5
5	From 95.5 to 97.3
10	From 96.4 to 97.9
20	From 97.2 to 98.1
50	From 97.6 to 98.4

For a given cost of materials, the efficiency will improve with the higher frequencies, and a transformer designed for a frequency of 25 would rarely have an efficiency higher than the lower limit given in the above table, while the higher figures apply mainly to transformers for use on 60-cycle circuits.

The highest efficiency of a lighting transformer usually occurs at about three-quarters of full load. Typical figures for a 5 k.v.a. lighting transformer for use on a 50-cycle circuit are given below.

Core loss = 46 watts.

Copper loss (full load) = 114 watts.

Calculated efficiency (100 per cent power factor):

At full load, 0.969.

At three-quarters full load, 0.9713.

At one-half full load, 0.9707.

At one-quarter full load, 0.9583.

FULL-LOAD EFFICIENCIES OF POWER TRANSFORMERS
FOR USE ON 66,000-VOLT CIRCUITS

(100 per cent power factor)

Output, k.v.a.	Efficiency, per cent.
400	From 97.3 to 97.8
800	From 97.7 to 98.2
1200	From 97.9 to 98.4
2000	From 98.1 to 98.7
2600	From 98.2 to 98.8

The manner in which the efficiency of large power transformers falls off with increase of voltage (involving loss of space taken up by insulation) is indicated by the following figures, which refer to 1000 k.v.a. single-phase units designed for use on 50-cycle circuits.

H.T. Voltage.	Full Load Efficiency (Approximate) Per cent.
22,000	98.8
33,000	98.7
44,000	98.5
66,000	98.3
88,000	98.0
110,000	97.8

The figures given below are actual test data showing the performance of some single-phase, oil-insulated, self-cooling, power transformers recently installed in a hydro-electric generating station in Canada:

Output.................... 400 k.v.a.

Frequency............... $f = 60$

Primary volts............. 2,200

Secondary volts........... 22,000

Core loss................. 1,760 watts

Full-load copper loss....... 3,550 watts

Exciting current, 2.15 per cent, of full-load current.

Temperature rise (by thermometer) after continuous full-load run, 36° C.

Efficiency on unity power factor load:

At 1.25 times full load.... 98.57 per cent

At full load............ 98.7

At three-quarters full load. 98.75

At one-half full load...... 98.65

At one-quarter full load... 98.0

It should be stated that the core loss in these transformers was exceptionally low, being only 0.44 per cent of the k.v.a. output. The core losses in modern transformers will usually lie between the limits stated below:

K.v.a. Output.	Volts.	Percentage Core Loss $= \dfrac{100 \times \text{core loss, watts.}}{\text{rated volt-ampere output}}$
500....	22,000	0.75 to 0.95
	66,000	1.0 to 1.2
1000....	22,000	0.6 to 0.7
	66,000	0.7 to 1.0
	110,000	0.8 to 1.15
2000....	22,000	0.5 to 0.65
	66,000	0.55 to 0.7
	110,000	0.7 to 0.95
4000....	66,000	0.5 to 0.6
	110,000	0.6 to 0.75

The core losses in small transformers for use on lighting circuits up to 2200 volts are usually less than 1 per cent for all sizes above 3 kw. They may be as low as 0.5 per cent in a 50 kw. distributing transformer, and as high as 2.5 per cent in a 1 kw. transformer.

The frequency, whether 25 or 60, does not greatly influence the customary allowance for core loss.

Efficiency when Power Factor of Load is Less than Unity. The total full-load losses (iron+copper) may be expressed as a percentage of the k.v.a. output. Assume that these losses are equal to a(k.v.a.). Then at any power factor, cos θ.

$$\text{Efficiency} = \frac{(\text{k.v.a.}) \cos \theta}{(\text{k.v.a.}) \cos \theta + a(\text{k.v.a.})},$$

$$= \frac{\cos \theta}{\cos \theta + a}.$$

Let η stand for the efficiency at unity power factor, then

$$\eta = \frac{1}{1+a},$$

and

$$a = \frac{1-\eta}{\eta},$$

whence the efficiency at any power factor, $\cos \theta$, is

$$\frac{\cos \theta}{\cos \theta + \left(\dfrac{1-\eta}{\eta}\right)}.$$

As an example, calculate the full-load efficiency of a transformer on a load of 0.75 power factor, given that the efficiency on unity power factor is 0.969.

The ratio of the total losses to the k.v.a. output is

$$a = \frac{1 - 0.969}{0.969} = 0.032$$

whence the efficiency at 0.75 power factor is

$$\frac{0.75}{0.75 + 0.032} = 0.959,$$

22. Temperature of Transformer Windings. Insulating materials such as cotton and paper, specially treated with insulating compounds or immersed in oil, may be subjected to a temperature up to, but not exceeding 105° C. The hottest spot of the winding cannot be reached by a thermometer, and it is therefore customary to add 15° C. to the temperature registered by a thermometer placed at the hottest accessible part of a transformer under test. The room temperature is frequently as high as 35° C. and the maximum permissible rise in temperature above that of the surrounding air may be arrived at as follows:

Permissible hottest spot temperature . 105°
Hottest spot correction   15
                                                    ——
Difference.   90
Assumed room temperature   35
                                                    ——
Difference (= permissible temperature
    rise)   55

Thus, under the worst conditions of heating, the permissible temperature rise should not exceed 55° C. when the measurements are made with a thermometer. A more reliable means of arriving at transformer temperatures is to calculate these from resistance measurements of the windings. Such measurements usually give somewhat higher temperatures than when thermometers are used, and a *hottest spot correction* of 10° C. is then generally recognized as sufficient. It should be noted, however, that room temperatures of 40° C. are not impossible, and it is therefore customary to limit the observed rise in temperature to 55° C. even when the resistance method of measuring temperatures is adopted.

Transformers are usually designed to withstand an overload of two hours' duration after having been in continuous operation under normal full-load conditions. Either of the following methods of rating is to be found in modern transformer specifications:

(1) The temperature rise not to exceed 40° C. on continuous operation at normal load, and 55° C. after an additional two hours' run on 25 per cent overload.

(2) The temperature rise not to exceed 35° C. on continuous operation at normal load, and 55° C. after an additional two hours' run on 50 per cent overload.

On account of the slow heating of the iron core, large oil-cooled transformers may require ten, or even twelve hours to attain the final temperature.

23. Heat Conductivity of Insulating Materials. Before discussing the means by which the heat is carried away from the external surface of the coils, it will be advisable to consider how the designer may predetermine approximately the difference in temperature between the hottest spot and the external surface of the windings. Calculations of internal temperatures cannot be made very accurately; but the nature of the problem is indicated by the following considerations:

Fig. 28 is supposed to represent a section through a very large flat plate, of thickness t, consisting of any homogeneous material. Assume a difference of temperature of $T_d = (T - T_0)$ ° C. to be maintained between

the two sides of the plate, and calculate the heat flow
(expressed in watts) through a portion of the plate of
area $w \times l$. The resistance offered by the material of
the plate to the passage of heat may be expressed in
thermal ohms, the thermal ohm being defined as the
thermal resistance which causes a drop of $1°$ C. per watt

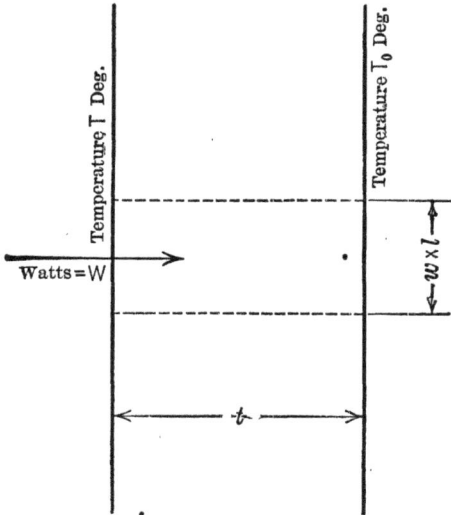

FIG. 28.—Diagram Illustrating Heat Flow through Flat Plate.

of heat flow; or, if R_h is the thermal resistance of the heat
path under consideration,

$$R_h = \frac{T_d}{W}, \quad \cdot \quad \cdot \quad \cdot \quad \cdot \quad \cdot \quad (19)$$

which permits of heat conduction problems being solved
by methods of calculation similar to those used in connection with the electric circuit.

Let k be the heat conductivity of the material, expressed in watts per inch cube per degree Centigrade
difference of temperature between opposite sides of the

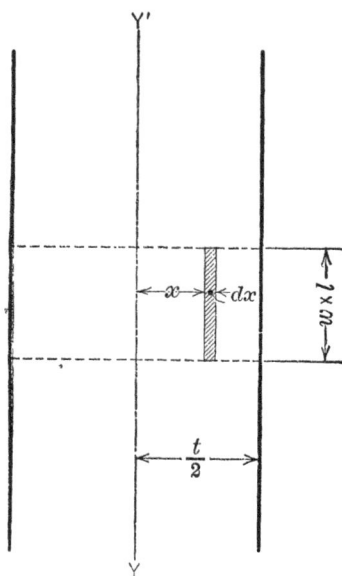

FIG. 29.—Heat Conductivity: Heat Generated Inside Plate.

cube, then the watts of heat flow crossing the area
$(w \times l)$ square inches, as indicated in Fig. 28 is

$$W = \left(\frac{w \times l}{t}k\right)T_d. \quad . \quad . \quad . \quad . \quad (20)$$

Fig. 29 illustrates a similar case, but the heat is now
supposed to be generated in the mass of the material
itself. We shall still consider the plate to be very large

relatively to the thickness, so that the heat flow from the center outward will be in the direction of the horizontal dotted lines. A uniformly distributed electric current of density Δ amperes per square inch is supposed to be flowing to or from the observer, and the highest temperature will be on the plane YY' passing through the center of the plate. Assuming this plate to be of copper with a resistivity of 0.84×10^{-6} ohms per inch cube at a temperature of about $80°$ C., the watts lost in a section of area $(x \times w)$ sq. in. and length l in. will be

$$W_x = (\Delta xw)^2 \times 0.84 \times 10^{-6} \times \frac{l}{xw}$$

$$= 0.84 \times 10^{-6} \Delta^2 wlx. \quad . \quad . \quad . \quad . \quad (21)$$

By adapting Formula (20) to this particular case, the difference of temperature between the two sides of a section dx in. thick is seen to be

$$dT_d = W_x \times \frac{dx}{wl \times k},$$

whence,

$$T_d = \frac{0.84 \times \Delta^2}{10^6 k} \int_0^{\frac{t}{2}} x\,dx$$

$$= \frac{0.84 \Delta^2 t^2}{8 \times 10^6 k} \text{ degrees Centigrade.} \quad . \quad . \quad (22)$$

The value of k for copper is about 10 watts per inch cube per degree Centigrade.

The problem of applying these principles to the practical case of a transformer coil is complicated by the fact

that the heat does not travel along parallel paths as in
the preceding examples, and, further, that the thermal
conductivity of the built-up coil depends upon the relative thickness of copper and insulating materials, a
relation which is usually different across the layers of
winding from what it is in a direction parallel to the
layers.

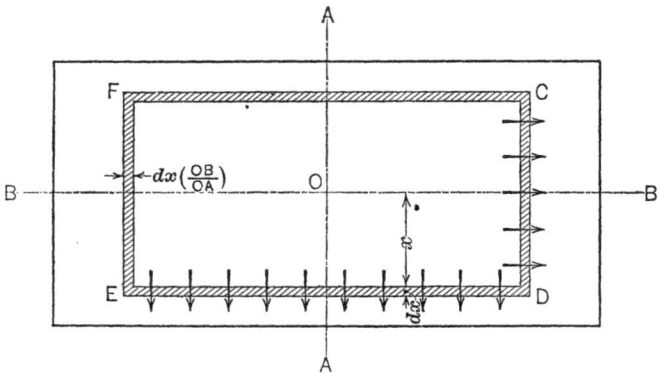

FIG. 30.—Diagram Illustrating Heat Paths in a Transformer Coil of
Rectangular Cross-section.

Fig. 30 represents a section through a transformer
coil wound with layers of wire in the direction OA;
the number of layers being such as to produce a total
depth of winding equal to twice OB. The whole of the
outside surface of this coil is supposed to be maintained
at a constant temperature by the surrounding oil or air.
In other words, it is assumed that there is a constant
difference of temperature of T_d degrees between the
hottest spot (supposed to be at the center O) and any
point on the surface of the coil.

The heat generated in the mass of material is thought of as traveling outward through the walls of successive imaginary spaces of rectangular section and length l (measured perpendicularly to the plane of the section shown in Fig. 30), as indicated in the figure, where $CDEF$ is the boundary of one of these imaginary spaces, the walls of which have a thickness dx in the direction OA, and a thickness $dx\left(\dfrac{OB}{OA}\right)$ in the direction OB.

According to Formula (19), we can say that the difference of temperature between the inner and outer boundaries of this imaginary wall is $dT_d = $ heat loss, in watts, occurring in the space $CDEF \times$ the thermal resistance of the boundary walls.

It is proposed to consider the heat flow through the portion of the boundary surface of which the area is $CDEF \times l$.

If W_x stands for the watts passing through this area, we can write

$$dT_d = W_x \times \frac{1}{\dfrac{2DElk_a}{dx} + \dfrac{2CDlk_b}{dx\left(\dfrac{OB}{OA}\right)}}.$$

which simplifies into

$$dT_d = W_x \frac{dx\left(\dfrac{OB}{OA}\right)}{2lx\left[k_a\left(\dfrac{OB}{OA}\right)^2 + k_b\right]} \quad \cdot \quad \cdot \quad \cdot \quad (23)$$

In order to calculate W_x it is necessary to know not only the current density, Δ, but also the space factor, or ratio of copper cross-section to total cross-section.

Let a stand for the thickness of copper per inch of total thickness of coil measured in the direction OA; and let b stand for a similar quantity measured in the direction OB; the space factor is then $(a \times b)$, and

$$W_x = \left[\Delta \times 2x \times 2x \left(\frac{OB}{OA} \right) \times ab \right]^2 0.84 \times 10^{-6} \cdot \frac{l}{2x \times 2x \left(\frac{OB}{OA} \right) ab}.$$

Inserting this value of W_x in (23), and making the necessary simplifications, we get

$$dT_d = \frac{0.84 \Delta^2 ab}{10^6 \left[k_a + k_b \left(\frac{OA}{OB} \right)^2 \right]} x dx,$$

whence, by integration between the limits $x = 0$ and $x = OA$,

$$T_d = \frac{0.84 \Delta^2 ab (OA)^2}{2 \times 10^6 \left[k_a + k_b \left(\frac{OA}{OB} \right)^2 \right]} \text{ deg. Cent. } . \quad (24)$$

Except for the obvious correction due to the introduction of the space factor (ab), the only difference between this formula and Formula (22) is that the thermal conductivity, instead of being k_a, as it would be if the heat flow were in the direction OA only, is replaced by the quantity in brackets in the denominator of Formula (24). This quantity may be thought of as a fictitious thermal conductivity in the direction OA, which, being greater than k_a, provides the necessary correction due to the fact that heat is being conducted away in the direction OB, thus reducing the difference of temperature between the points O and A.

Calculation of k_a and k_b.

Let k_c and k_i, respectively, stand for the thermal conductivity of copper and insulating materials as used in transformer construction. The numerical values of these quantities, expressed in watts per inch cube per degree Centigrade, are $k_c = 10$ and $k_i = 0.0033$. It follows that

$$k_a = \frac{1}{\dfrac{a}{k_c} + \dfrac{(1-a)}{k_i}} = \frac{10}{a + 3000(1-a)}, \quad \cdot \quad \cdot \quad (25)$$

and similarly,

$$k_b = \frac{10}{b + 3000(1-b)}, \quad \cdot \quad \cdot \quad \cdot \quad (26)$$

where a and b are the thickness of copper per inch of coil in the directions OA and OB, respectively, as previously defined.

Example. Suppose a transformer coil to be wound with 0.25×0.25 in. square copper wire insulated with cotton 0.01 in. thick, and provided with extra insulation of 0.008 in. fullerboard between layers. There are twelve layers of wire and seven wires per layer. Assume the current density to be 1400 amperes per square inch, and calculate the hottest spot temperature if the outside surface of the coil is maintained at 75° C.

$$a = \frac{0.25}{0.27} = 0.926; \quad b = \frac{0.25}{0.278} = 0.9; \quad \text{whence space factor}$$
$(ab) = 0.833$.

By Formulas (25) and (26), $k_a = 0.0448$, and $k_b = 0.0332$; $OA = 3.5 \times 0.27 = 0.945$, and $OB = 6 \times 0.278 = 1.67$ in.

By Formula (24),

$$T_d = \frac{0.84(1400)^2 \times 0.833(0.945)^2}{2 \times 10^6 \left[0.0448 + 0.0332 \left(\frac{0.945}{1.67} \right)^2 \right]} = 11° \text{ Cent.,}$$

and the hottest spot temperature $= 75 + 11 = 86°$ C.

24. Cooling Transformers by Air Blast. Before the advantages of oil insulation had been realized, transformers were frequently enclosed in watertight cases, the metal of these cases being separated from the hot parts of the transformer by a layer of still air. This resulted either in high temperatures or in small kilowatts output per pound of material. Air insulation is still used in some designs of large transformers for pressures up to about 33,000 volts; but efficient cooling is obtained by forcing the air around the windings and through ducts provided not only between the coils, but also between the coils and core, and between sections of the core itself.

Since all the heat losses which are not radiated from the surface of the transformer case must be carried away by the air blast, it is a simple matter to calculate the weight (or volume) of air required to carry away these losses with a given average increase in temperature of outgoing over ingoing air.

A cubic foot of air per minute, at ordinary atmospheric pressures, will carry away heat at the rate of about 0.6 watt for every degree Centigrade increase of temperature. Thus, if the difference of temperature between outgoing and ingoing air is 10° C., the quantity of air which must pass through the transformer for every kilowatt of total loss that is not radiated from the surface of the case, is

$$Q = \frac{1000}{0.6 \times 10} = 166 \text{ cu. ft. per minute.}$$

If the average increase in temperature of the air is from 10 to 15° C., the actual surface temperature rise of the windings may be from 40 to 50° C.; the exact figure being difficult to calculate since it will depend upon the size and arrangement of the air ducts. The temperature of the coils is influenced not only by the velocity of the air over the heated surfaces, but also by the amount of the total air supply which comes into intimate contact with these surfaces. With air passages about $\frac{1}{2}$ in. wide, and an average air velocity through the ducts ranging from 300 to 600 ft. per minute, the temperature rise of the coil. surfaces will usually be from four to eight times the rise in temperature of the circulating air. Thus, although it is not possible to predetermine the exact quantity of air necessary to maintain the transformer windings at a safe temperature, this may be expressed approximately as:

$$\left.\begin{array}{l}\text{Cubic feet of air per minute}\\\text{for } 50^\circ \text{ C. temperature rise}\\\text{of coil surface}\end{array}\right\} = \frac{W_t - W_r}{0.6 \times \frac{50}{6}},$$

$$= 0.2(W_t - W_r), \quad . \quad (27)$$

where $W_t =$ total watts lost in transformer; and
$\quad\quad W_r =$ portion of total loss dissipated from surface of tank.

The latter quantity may be estimated by assuming the temperature of the case to be about 10° C. higher than that of the surrounding air, and calculating the watts radiated from the case with the aid of the data in the succeeding article.

Assuming W_r to be 25 per cent of W_t, the Formula (27) indicates that about 150 cu. ft. of air per minute per kilowatt of total losses would be necessary to limit the temperature rise of the coils to 50° C. With poorly designed transformers, and also in the case of small units, the amount of air required may be appreciably greater.

It is true that, in turbo-generators, an allowance of 100 cu. ft. per minute per kilowatt of total losses, is generally sufficient to limit the temperature rise to about 50° C.; but, owing to the churning of the air due to the rotation of the rotor, it would seem that the necessary supply of air is smaller for turbo-generators than for transformers.

Filtered air is necessary in connection with air-blast cooling; otherwise the ventilating ducts are liable to become choked up with dirt, and high temperatures will result. Wet air filters are very satisfactory and desirable, provided the amount of moisture in the air passing through the transformers is not sufficient to cause a deposit of water particles on the coils. Air containing from 1 to 3 per cent of free water in suspension is a much more effective cooling medium than dry air. It would probably be inadvisable to use anything but dry air in contact with extra-high voltage apparatus; but transformers for very high pressures are not designed for air-blast cooling.*

25. Oil-immersed Transformers--Self Cooling. The natural circulation of the oil as it rises from the heated surfaces of the core and windings, and flows downward near the sides of the containing tank, will lead to a temperature distribution generally as indicated in Fig. 31. The temperature of the oil at the hottest part (close to the windings at the top of the transformer) will be somewhat higher than the maximum temperature of the tank, which, however, will be hotter in the neighborhood of the oil level than at other parts of its surface. The average temperature of the cooling surface in contact with the air bears some relation to the highest oil temperature, and, since this relation does not vary greatly with different designs of transformer, or case, a curve such

* Some useful data on the relative cooling effects of moist and dry air, together with test figures relating to a 12-kw. air-cooled transformer, will be found in Mr. F. J. Teago's paper " Experiments on Air-blast Cooling of Transformers," in the *Jour. Inst. E. E.*, May 1, 1914 ,Vol. 52, page 563.

as Fig. 32 may be used for calculating the approximate
tank are a necessary to prevent excessive oil temperatures.

The oil temperature referred to in Fig. 32 is the difference in degrees Centigrade between the temperature
of the hottest part of the oil and the air outside the tank.

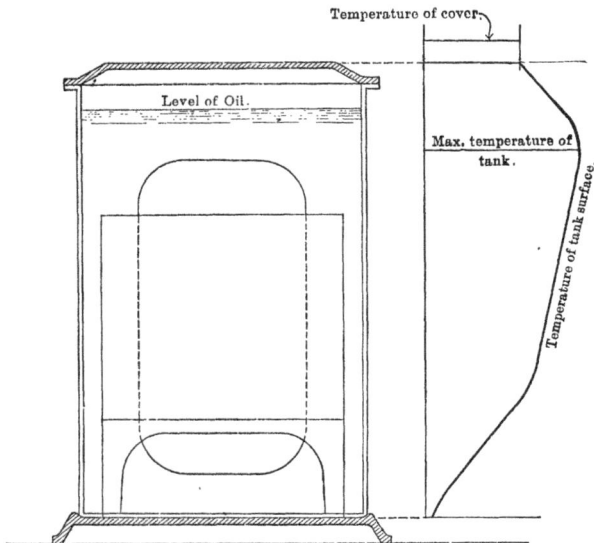

FIG. 31.—Distribution of Temperature with Transformer Immersed in
Oil.

This will be somewhat greater than the temperature rise
of any portion of the transformer case; but the curve
indicates the (approximate) number of watts that can
be dissipated—by radiation and air currents—per square
inch of tank surface. The curve is based on average
figures obtained from tests on tanks with *smooth surfaces*

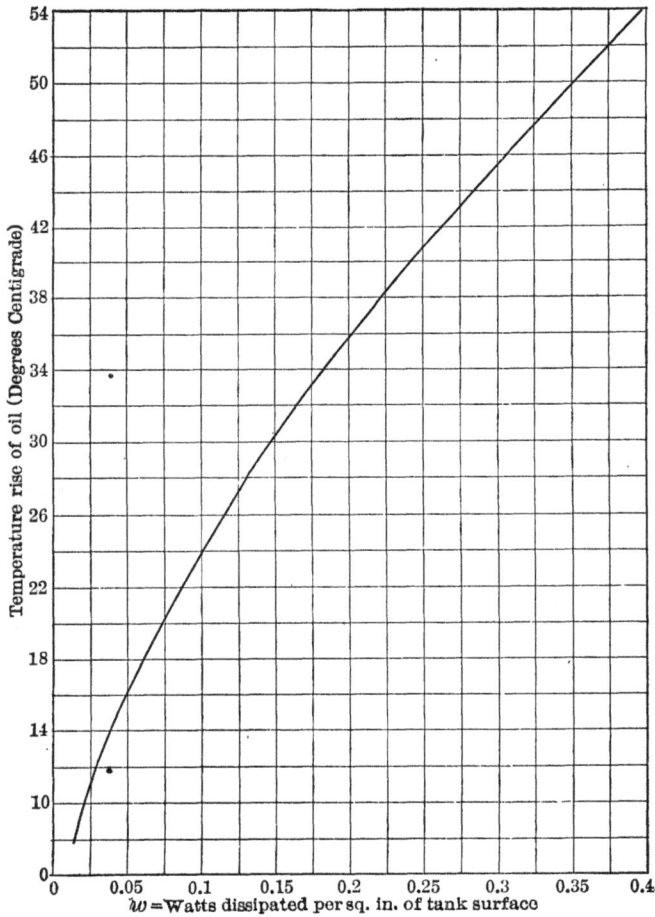

FIG. 32.—Curve for Calculating Cooling Area of Transformer Tanks.

(not corrugated), the surface considered being the total area of the (vertical) sides *plus* one-half the area of the lid. The cooling effect of the bottom of the tank is practically negligible, and is not to be included in the calculations.

Example. What will be the probable maximum temperature rise of the oil in a self-cooling transformer with a total loss of 1200 watts, the tank—of sheet-iron without corrugations—measuring 2 ft. × 2 ft. × 3.5 ft. high?

The surface for use in the calculations is $S = (3.5 \times 8) + 2 = 30$ sq. ft., whence

$$w = \frac{1200}{30 \times 144} = 0.278,$$

which, according to Fig. 32, indicates a 43° C. rise of temperature for the oil.

The temperature of the windings at the hottest part of the surface in contact with the oil might be from 5 to 10° C. higher than the maximum oil temperature as measured by thermometer. Assume this to be 7° C. Assume also that the room temperature is 35° C. and that the difference of temperature (T_d) between the coil surface and the hottest spot of the windings—as calculated by the method explained in Art. 23—is 13° C. Then the hottest spot temperature in the transformer under consideration would be about $35 + 43 + 7 + 13 = 98$° C.

26. Effect of Corrugations in Vertical Sides of Containing Tank. The cooling surface in contact with the air may be increased by using corrugated sheet-iron tanks in place of tanks with smooth sides. It must not, however, be supposed that the temperature reduction will be proportional to the increase of tank surface provided in this manner; the watts radiated per square inch of surface of a tank with corrugated sides will always be appreciably less than when the tank has smooth sides. Not only is the surface near the bottom of the corrugations less effective in radiating heat than the outside portions; but the depth and pitch of the corrugations will affect the (downward) rate of flow of the oil on the inside of the tank, and the (upward) convection currents of air on the outside.

It is practically impossible to develop formulas which will take accurate account of all the factors involved, and recourse must therefore be had to empirical formulas based on available test data together with such reasonable assumptions as may be necessary to render them suitable for general application.

If λ is the pitch of the corrugations, measured on the outside of the tank, and l is the surface width of material per pitch (see the sketch in Fig. 33), the ratio of the actual tank surface to the surface of a tank without corrugations is $\frac{l}{\lambda}$. The heat dissipation will not be in this proportion because, although the cooling effect will increase as l is made larger relatively to λ, the additional surface becomes less and less effective in radiating heat as the depth of the corrugations increases without a corresponding increase in the pitch. It is convenient to think of the surface of an equivalent smooth tank which will give the same temperature rise of the oil as will be obtained with the actual tank.

If we apply a correction to the actual pitch, λ, and obtain an equivalent pitch, λ_e, the ratio $k = \dfrac{\lambda_e}{\lambda}$ is a factor by which the tank surface (neglecting corru-

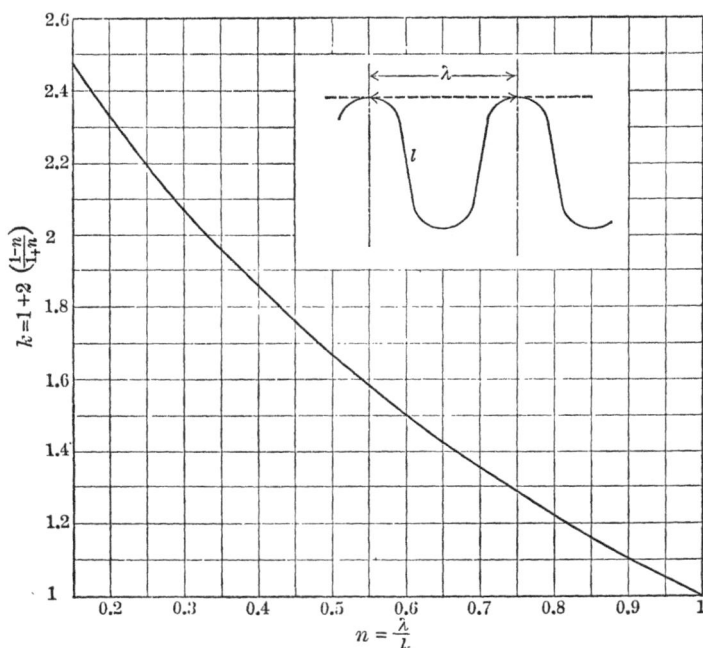

FIG. 33.—Curve Giving Factor k for Calculating Equivalent Cooling Surface of Tanks with Corrugated Sides.

gations) must be multiplied in order to obtain the *equivalent* or *effective* surface. If all portions of the added surface were equally effective in radiating heat, no correcting factor would be required, and the equivalent pitch would be obtained by adding to λ the quantity $(l-\lambda)$; but since a modifying factor is needed, the writer proposes the formula

$$\lambda_e=\lambda+(l-\lambda)\left(\frac{2\lambda}{l+\lambda}\right), \quad . \quad . \quad . \quad . \quad (28)$$

wherein the additional surface provided by the corrugations is reduced in the ratio $\frac{2\lambda}{l+\lambda}$ which becomes unity when $l=\lambda$. A modifying factor of this form not only seems reasonable on theoretical grounds, but it is required in an empirical formula based on available experimental data. It follows that

$$k=\frac{\lambda_e}{\lambda}=1+\frac{2(l-\lambda)}{l+\lambda}, \quad . \quad . \quad . \quad (29)$$

or, if $\frac{\lambda}{l}=n$,

$$k=1+2\left(\frac{1-n}{1+n}\right). \quad . \quad . \quad . \quad . \quad (30)$$

Values of k, as obtained from this formula for different values of n, may be read off the curve Fig. 33.

Example. What would have been the temperature rise of the oil if, instead of the smooth-side tank of the preceding example (Art. 25), a tank of the same external dimensions had been provided with corrugations 2 in. deep, spaced $1\frac{1}{4}$ in. apart?

The approximate value of l is $1.25+4=5.25$ in. Whence $n=\frac{1.25}{5.25}=0.238$; and from the curve, Fig. 30, $k=2.23$.

The equivalent tank surface is $S = (3.5 \times 8 \times 2.23) + 2$
$= 64.5$ sq. in., whence

$$w = \frac{1200}{64.5 \times 144} = 0.129,$$

which, according to Fig. 32, indicates a $27°$ C. rise of temperature, as compared with $43°$ C. with the smooth-surface tank of the same outside dimensions.

27. Effect of Overloads on Transformer Temperatures. Since the curve of Fig. 32 is not a straight line, it follows that the watts dissipated per square inch of tank surface are not directly proportional to the difference between the oil, and room, temperatures. The approximate relation, according to this curve, is

$$\text{Temperature rise} = \text{constant} \times w^{0.6}, \quad . \quad . \quad (31)$$

which may be used for calculating the temperature rise of a self-cooling oil-immersed transformer when the temperature rise under given conditions of loading is known.

Example. Given the following particulars relating to a transformer:

> Core loss = 100 watts,
> Copper loss (full load) = 200,
> Final temp. rise (full load) of the oil = $35°$ C.

Calculate the final temperature rise after a continuous run at 20 per cent overload.

For an increase of 20 per cent in the load, the copper

loss is $200 \times (1.2)^2 = 288$ watts; whence, according to Formula (31):

$$\text{Temperature rise} = 35 \times \left(\frac{288+100}{200+100}\right)^{0.6} = 41° \text{ C. approx.}$$

The calculation of temperature rise resulting from an overload of short duration is not so simple. It is necessary to take account of the specific heat of the materials, especially the oil, because the heat units absorbed by the materials have not to be radiated from the tank surface, and the calculated temperature rise would be too high if this item were neglected.

The specific heat of a substance is the number of calories required to raise the temperature of 1 gram 1° C., the specific heat of water being taken as unity.

The specific heat of copper is 0.093, and for an average quality of transformer oil, it is 0.32.

One gram-calorie (*i.e.*, the heat necessary to raise the temperature of 1 gram of water 1° C.) $= 4.183$ joules (or watt-seconds). Also, 1 lb. $= 453.6$ grams. It follows that the amount of energy in watt-seconds necessary to raise M_c pounds of copper $T°$ C. is

$$\text{Watts} \times \text{time in seconds} = 4.183 \times 0.093 \times 453.6 \, M_c T$$
$$= 177 \, M_c T \text{ (for copper)}.$$

Similarly, if we put M_o for the weight of oil, in pounds, and replace the figure 0.093 by 0.32, we get

$$\text{Watts} \times \text{time in seconds} = 610 \, M_o T \text{ (for oil)}.$$

In the case of an overload after the transformer has been operating a considerable length of time on normal full load, all the additional losses occur in the copper coils, and it is generally permissible to 'neglect the heat absorption by the iron core. We shall, therefore, assume that the additional heat units which are not absorbed by the copper pass into the oil, and that the balance, which is not needed to heat up the oil, must be dissipated by radiation and convection from the sides of the containing tank. It will greatly simplify the calculations if we further assume that the watts dissipated per square inch of tank surface per degree difference of temperature are constant over the range of temperature involved in the problem. (By estimating the average temperature rise, and finding w on the curve, Fig. 32, a suitable value for the quantity $\dfrac{w}{T}$ may be selected.)

If W_t = total watts lost (iron + copper), the total energy loss in the interval of time dt second is $W_t dt$.

If the increase of temperature during this interval of time is dx degree Centigrade, the heat units absorbed by the copper coils and the oil are $K_s dx$, where $K_s = (177 M_c + 610 M_0)$.*

The difference between these two quantities represents the number of joules, or watt-seconds, of energy to be

---

* In order to simplify the calculations, it has been (incorrectly) assumed that the temperature rise of the copper is the same as that of the oil. It will, of course, be somewhat greater; but since the heat absorbed by the copper is small compared with that absorbed by the oil, this assumption will not lead to an error of appreciable magnitude.

radiated from the tank surface during the interval of time dt second; whence,

$$W_t dt - K_s dx = K_r x dt, \quad . \quad . \quad . \quad . \quad (32)$$

where K_r = tank surface in square inches × radiation coefficient, in watts per square inch per $1°$ C. rise, and x = the initial oil temperature rise (which has been increased by the amount dx).

Equation (32) may be put in the form

$$\frac{dt}{dx} = \frac{K_s}{W_t - K_r x}.$$

The limits for x are the initial oil temperature T_0 and the final oil temperature T_t, which is reached at the end of the time t. Therefore,

$$t = \int_{T_0}^{T_t} \frac{K_s}{W_t - K_r x} dx$$

$$= \frac{K_s}{K_r} \log_e \left(\frac{\dfrac{W_t}{K_r} - T_0}{\dfrac{W_t}{K_r} - T_t} \right). \quad . \quad . \quad . \quad (33)$$

If time is expressed in minutes, and common logs. are used, we have,

$$t_m = \frac{K_s}{26 K_r} \log_{10} \left(\frac{\dfrac{W_t}{K_r} - T_0}{\dfrac{W_t}{K_r} - T_1} \right) \text{ minutes.} \quad . \quad . \quad (34)$$

In order to facilitate the use of this formula, the meaning of the symbols is repeated below:

W_t = total watts lost (iron + copper);

K_r = $S \times$ radiation coefficient expressed in watts per square inch per $1°$ C. rise of temperature of the oil;

    where S = tank surface in square inches, as defined in Art. 25, corrected if necessary for corrugations (Art. 26).

K_s = $177 M_c + 610 M_0$;

    where M_c = weight of copper (pounds);

    and M_0 = weight of oil (pounds);

T_0 = initial temperature of oil (degrees C.);

T_t = temperature of oil (degrees C.) after the overload (producing the total losses W_t) has been on for t_m minutes.

Example. Using the data of the preceding example, the full-load conditions are:

        Core loss   = 100 watts;
        Copper loss   = 200 watts;
        Temperature rise = $35°$ C.

Referring to Fig. 29, the value for w for a temperature rise of $35°$ is 0.193, from which it follows that the effective tank surface is $S = \dfrac{100 + 200}{0.193} = 1550$ sq. in.

Given the additional data:

        Weight of copper = 65 lb.,
        Weight of oil   = 140 lb.

calculate the time required to raise the oil from $T_0 = 35°$ C. to $T_t = 45°$ C. on an overload of 50 per cent.

The copper loss is now $200 \times (1.5)^2 = 450$ watts, whence

$$W_t = 100 + 450 = 550 \text{ watts.}$$

The cooling coefficient (from curve, Fig. 32), for an average temperature rise of $\dfrac{35+45}{2} = 40°$ C., is $\dfrac{0.242}{40}$ $= 0.00606$, whence,

$$K_r = 1550 \times 0.00606 = 9.4;$$

$$K_s = (177 \times 65) + (610 \times 140) = 97,000;$$

and, by Formula (34),

$$t_m = \frac{97,000}{26 \times 9.4} \log\left(\frac{\dfrac{550}{9.4} - 35}{\dfrac{550}{9.4} - 45}\right) = 95.5 \text{ minutes.}$$

28. Self-cooling Transformers for Large Outputs. The best way to cool large transformers is to provide them with pipe coils through which cold water is circulated, or, alternatively, to force the oil through the ducts and provide means for cooling the circulating oil outside the transformer case. When such methods cannot be adopted—as in most outdoor installations and other sub-stations without the necessary machinery and attendants—the heat from self-cooling transformers of large size is dissipated by providing additional cooling surface in the form of tubes, or flat tanks of small volume and large external surface, connected to the outside of a central containing tank. Unless test data are available in connection with the particular design adopted, judgment is needed to determine the effective cooling surface (see Art. 26) in order that the curve of Fig. 32—or such cooling data as may be available for smooth-surface tanks—may be used for calculating the probable temperature rise.

In the tubular type of transformer tank which is provided with external vertical tubes connecting the bottom of the tank to the level, near the oil surface, where the temperature is highest (as roughly illustrated by Fig. 34), the tubes should be of fairly large diameter with sufficient distance between them to allow free circulation of the air and efficient radiation. It is not economical to use a very large number of small tubes closely spaced with a view to obtaining a large cooling surface, because the extra surface obtained by such means is not as effective as when wider spacing is used.

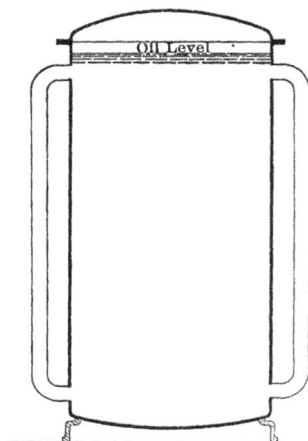

FIG. 34.--Transformer Case with Tubes to Provide Additional Cooling Surface.

If the added pipe surface, A_p, is 1.5 times the tank surface, A_t, without the pipes, the effective cooling surface will be about $S = (A_t + A_p) \times 0.9$; but, with a greatly increased surface obtained by reducing the spacing between the pipes, the correction factor might be very much smaller than 0.9.

29. Water-cooled Transformers. The cooling coil should be constructed preferably of seamless copper tube about $1\frac{1}{4}$ in. diameter, placed near the top of the tank, but below the surface of the oil. If water is passed through the coil, heat will be carried away at the rate of 1000 watts for every $3\frac{3}{4}$ gals. flowing per minute when the difference of temperature between the outgoing and ingoing water is $1°$ C. Allowing 0.25 gal. per minute, per kilowatt, the average temperature rise of the water will be $\dfrac{3.75}{0.25} = 15°$ C. The temperature rise of the oil is considerably greater than this: it will depend upon the area of the coil in contact with the oil and the condition of the inside surface, which may become coated with scale. An allowance of 1 sq. in. of coil surface per watt is customary; but the rate at which heat is transferred from the oil to the water may be from 2 to $2\frac{1}{2}$ times as great when the pipes are new than after they have become coated with scale. It may, therefore, be necessary to clean them out with acid at regular intervals, if the danger of high oil temperatures is to be avoided.

Example. Calculate the coil surface and the quantity of water required for a transformer with total losses amounting to 6 kw., of which it is estimated that 2 kw. will be dissipated from the outside of the tank. Surface of cooling coil $= 6000 - 2000 = 4000$ sq. in.

Assuming a diameter of $1\frac{1}{4}$ in., the length of tube in the coil will have to be $\dfrac{4000}{12 \times 1.25 \times \pi} = 85$ ft.

The approximate quantity of water required will be $0.25 \times 4 = 1$ gal. per minute.

30. Transformers Cooled by Forced Oil Circulation. The transformer and case are specially designed so that the oil may be forced (by means of an external pump) through the spaces provided between the coils and between the sections of the iron core. The ducts may be narrower than when the cooling is by natural circulation of the oil. The capacity of the oil pump may be estimated by allowing a rate of flow of oil through the ducts ranging from 20 to 30 ft. per minute.

It is not essential that the oil be cooled outside the transformer case; in some modern transformers, the containing tank proper is surrounded by an outer case, and the space between these two shells contains the cooling coils through which water is circulated. These coils, instead of being confined to the upper portion of the transformer case, as when water cooling is used without forced oil circulation, may occupy the whole of the space between the inner and outer shells of the containing tank. The oil circulation is obtained by forcing the oil up through the inner chamber and downward in the space surrounding the water cooling-coils.

Such systems of artificial circulation of both oil and water are very effective in connection with units of large output; but they could not be applied economically to medium-sized or small units.

CHAPTER IV

31. Magnetic Leakage. Assuming the voltage applied to the terminals of a transformer to remain constant, it follows that the flux linkages necessary to produce the required back e.m.f. can readily be calculated. The (vectorial) difference between the applied volts and the induced volts must always be exactly equal to the ohmic drop of pressure in the primary winding. Thus, the total primary flux linkages (which may include leakage lines) must be such as to induce a back e.m.f. very nearly equal to the applied e.m.f.—the primary IR drop being comparatively small.

When the secondary is open-circuited, practically all the flux linking with the primary turns links also with the secondary turns; but when the transformer is loaded, the m.m.f. due to the current in the secondary winding has a tendency to modify the flux distribution, the action being briefly as follows:

The magnetomotive force due to a current I_s flowing in the secondary coils would have an immediate effect on the flux in the iron core if it were not for the fact that the slightest tendency to change the number of flux lines through the primary coils instantly causes the primary current to rise to a value I_p such that the resultant

107

ampere turns $(I_p T_p - I_s T_s)$ will produce the exact amount of flux required to develop the necessary back e.m.f. in the primary winding. Thus, the total amount of flux linking with the primary turns will not change appreciably when current is drawn from the secondary terminals; but the secondary m.m.f.—together with an exactly equal but opposite primary magnetizing effect—will cause some of the flux which previously passed through the secondary core to " spill over " and avoid some, or all, of the secondary turns. This reduces the secondary volts by an amount exceeding what can be accounted for by the ohmic resistance of the windings.

Although it is possible to think of a leakage field set up by the secondary ampere turns independently of that set up by the primary ampere turns, these imaginary flux components must be superimposed on the main flux common to both primary and secondary in order that the resultant magnetic flux distribution under load may be realized. The leakage flux is caused by the combined action of primary and secondary ampere turns, and it is incorrect, and sometimes misleading, to think of the secondary leakage reactance of a transformer as if it were distinct from primary reactance, and due to a particular set of flux lines created by the secondary current. In order to obtain a physical conception of magnetic leakage in transformers it is much better to assume that the secondary of an ordinary transformer has no *self*-inductance, and that the loss of pressure (other than IR drop) which occurs under load is caused by the secondary ampere turns diverting a certain amount of magnetic flux which, although it still links with the primary turns, now follows certain leakage paths instead of passing through the core under the secondary coils.

32. Effect of Magnetic Leakage on Voltage Regulation. The regulation of a transformer may be defined as the percentage increase of secondary terminal voltage when the load is disconnected (primary impressed voltage and frequency remaining unaltered).

The connection between magnetic leakage and voltage regulation will be studied by considering the simplest possible cases, and noting the difference in secondary flux-linkages under loaded and open-circuited conditions. The amount of the leakage flux in proportion to the useful flux will purposely be greatly exaggerated, and, in order to eliminate unessential considerations, the following assumptions will be made:

(1) The magnetizing component of the primary current will be considered negligible relatively to the total current, and will not be shown in the diagrams.

(2) The voltage drop due to ohmic resistance of both primary and secondary windings will be neglected.

(3) The primary and secondary windings will be symmetrically placed and will consist of the same number of turns.

(4) One flux line—as shown in the diagrams—linking with one turn of winding will generate one volt.

In Fig. 35, both primary and secondary coils consist of one turn of wire wound close around the core: a current I_s is drawn from the secondary on a load of power factor $\cos\theta$, causing a current I_1—exactly equal but opposite to I_s—to flow in the primary coil, the result being the leakage flux as represented by the four dotted lines.

The secondary voltage, $E_s = 2$ volts, is due to the two flux lines which link both with the primary and secondary

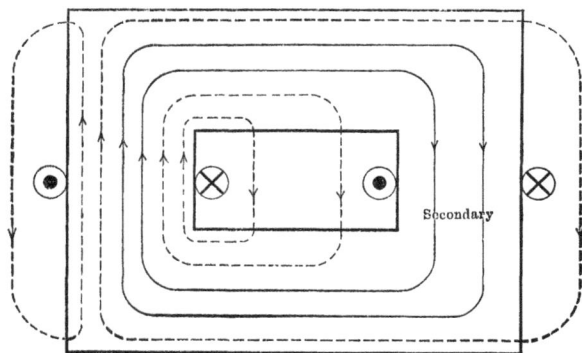

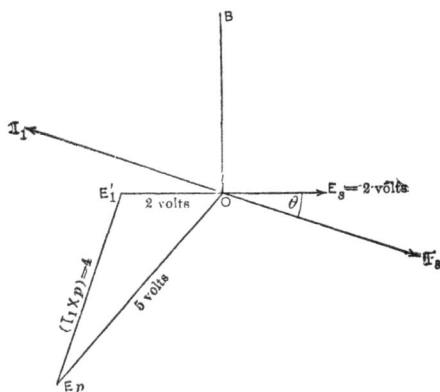

FIG. 35.—Magnetic Leakage: Thickness of Coils Considered Negligible.

coils.   The phase of this component of the total flux is, therefore, $90°$ in advance of E_s as indicated by the line OB in the vector diagram.

In order to calculate the necessary primary impressed e.m.f., we have as one component OE'_1 exactly equal but opposite to OE_s because the flux OB will induce in the primary coil a voltage exactly equal to E_s in the secondary. The other component is $E'_1E_p = 4$ volts, equal but opposite to the counter e.m.f. which, being due to the four leakage lines created by the current I_1, will lag 90° in phase behind OI_1. The resultant is OE_p which scales 5 volts.*

When load is thrown off the transformer there will be five lines linking with the primary which, since there is now no secondary m.m.f. to produce leakage flux, will pass through the iron core and link with the secondary. The secondary voltage on open circuit will, therefore, be $E_p = 5$ volts, and the percentage regulation is

$$100 \times \frac{E_p - E_s}{E_s} = 100 \times \frac{5-2}{2} = 150.$$

In Fig. 36, a departure is made from the extreme simplicity of the preceding case in order to illustrate the effect of leakage lines passing not only entirely outside the windings, but also through the thickness of the coils, as must always happen in practical transformers where the coils occupy an appreciable amount of space.

Each winding now consists of two turns, with an air space between the turns through which leakage flux--

* The reason why the six flux lines shown in the figure as linking with the primary coil do not generate 6 volts is, of course, due to the fact that these flux lines are not all in the same phase; the *resultant* or actual flux in the core under the primary coil is 5 lines, as indicated by the vector diagram. The actual amount of flux passing any given cross-section of the core must be thought of as the (vectorial) addition of the flux lines shown in the sketch at that particular section.

represented in Fig. 36 by one dotted line—is supposed to pass. The single flux line, linking with both the primary

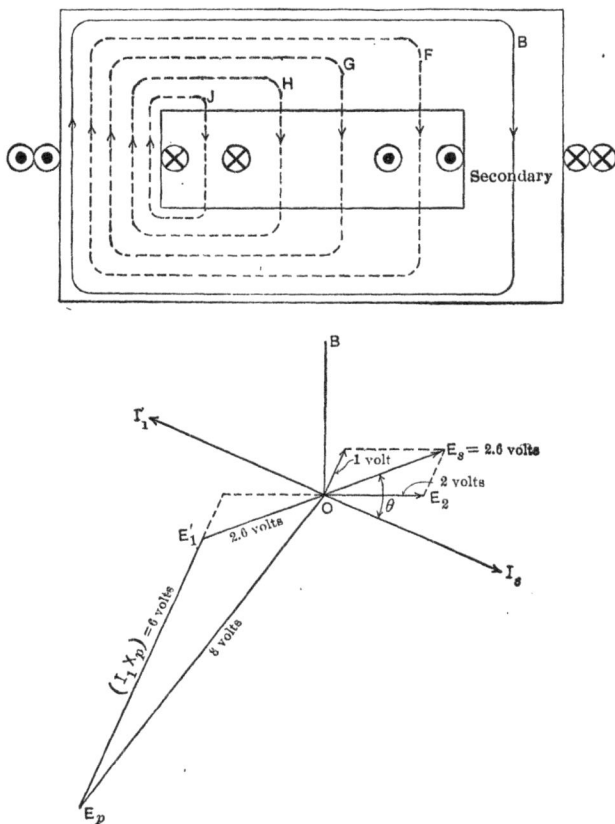

FIG. 36.—Magnetic Leakage: Thickness of Coils Appreciable.

and secondary windings, generates the e.m.f. component $E_9 = 2$ volts. The leakage flux line marked F links with only óne turn of the secondary, and therefore generates one volt lagging 90° in phase behind the primary current I_1. The total secondary voltage is E_s which scales 2.6 volts; the balancing component in the primary being E_1. It should be particularly noted that this balancing component does not account for the full effect of the two flux lines B and F linking with the primary, because, while the flux line F links with only one secondary turn, it links with two primary turns. The voltage component OE'_1 in the primary may, therefore, be thought of as due to the flux lines B and J, leaving for the remaining component of the impressed e.m.f., $E'_1 E_p = 6$ volts (leading OI_1 by 90°) which may be considered as caused by the three lines F, H, and G. In other words, the reactive drop $(I_1 X_p)$ depends upon the *difference* between the primary and secondary flux-linkages of the stray magnetic field set up by the combined action of the secondary current I_s and the balancing component I_1 of the total primary current. (In this case I_1 is the total primary current, since the magnetizing component is neglected.) The leakage flux-linkages are as follows:

With the primary turns:

$$J \times 1 = 1 \text{ volt,}$$
$$H \times 2 = 2 \text{ volts,}$$
$$G \times 2 = 2 \text{ volts,}$$
$$F \times 2 = 2 \text{ volts}$$

7 volts

With the secondary turns:

$$F \times 1 = 1 \text{ volt}$$

giving a difference of   6 volts

This is the vector $(I_1 X_p)$. When applying this rule to actual transformers in which the ratio of turns $\dfrac{T_p}{T_s}$ is not unity, the proper correction must be made (as explained later) when calculating the equivalent e.m.f. component in the primary circuit.

To obtain the regulation in the case of Fig. 36, we have $E_p = 8$ volts and $E_s = 2.6$ volts when the transformer is loaded. When the load is thrown off, there will be four flux lines linking with both primary and secondary producing 8 volts in each winding. The regulation is therefore,

$$100 \times \frac{8 - 2.6}{2.6} = 208 \text{ per cent.}$$

33. Experimental Determination of the Leakage Reactance of a Transformer. Although these articles are written from the viewpoint of the designer, who must predetermine the performance of the apparatus he is designing, a useful purpose will be served by considering how the leakage reactance of an actual transformer may be determined on test. The purpose referred to is the clearing up of any vagueness and consequent inaccuracy that may exist in the mind of the reader, due largely— in the writer's opinion—to the common, but unnecessary if not misleading assumption, that the secondary has self induction.*

---

* The assumption usually made in text books is that the secondary self-induction (i.e., the flux produced by the secondary current, and linking with the secondary turns) is equal to the primary leakage self-induction.

The diagram, Fig. 37, shows the secondary of a transformer short-circuited through an ammeter, A, of negligible resistance. The impressed primary voltage E_z, of the frequency for which the transformer is designed, is adjusted until the secondary current I_s is indicated by the ammeter. If the number of turns in the primary and secondary are T_p and T_s respectively, the primary current will be $I_1 = I_s\left(\dfrac{T_s}{T_p}\right)$ because, the amount of flux in the core being very small, the mag-

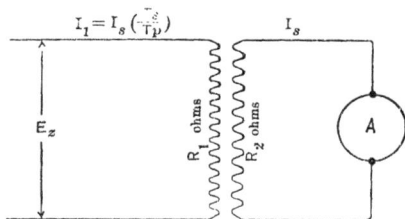

FIG. 37.—Diagram of Short-circuited Transformer.

netizing component of the primary current may be neglected.

The measured resistances R_1 and R_2 of the primary and secondary coils being known, the vector diagram Fig. 38, can be constructed.

The volts induced in the secondary are OE_2 (equal to I_sR_2) in phase with the current I_s. The balancing component in the primary winding is OE'_1 $\left[\text{equal to } E_2\left(\dfrac{T_p}{T_s}\right)\right]$ in phase with the primary current I_1. Another component in phase with this current is E'_1P (equal to I_1R_1). Since the total impressed voltage has the known value E_z, we can describe an arc of. circle of radius OE_z from the point O as a center. By erecting a perpendicular to OI_1 at the point P, the point E_z is determined, and E_zP is the loss of pressure caused by magnetic leakage. The vector OP may be thought of as the product of the primary current I_1,

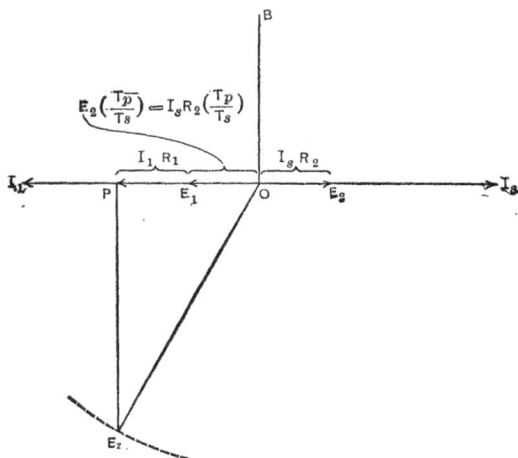

FIG. 38.—Vector Diagram of Short-circuited Transformer.

and an equivalent primary resistance R_p, which assumes the secondary resistance to be zero, but the primary resistance to be increased by an amount equivalent to the actual secondary resistance. Thus,

$$I_1R_p = I_1R_1 + I_sR_2\left(\frac{T_p}{T_s}\right),$$

but

$$I_s = I_1\left(\frac{T_p}{T_s}\right),$$

whence,

$$R_p = R_1 + R_2\left(\frac{T_p}{T_s}\right)^2. \quad \cdots \quad (35)$$

In order to get an expression for the transformer leakage reactance (X_p) in terms of the test data, we can write,

$$E_z = I_1\sqrt{R_p{}^2 + X_p{}^2}:$$

whence

$$X_p = \sqrt{\frac{E_z{}^2}{I_1{}^2} - R_p{}^2}.$$

This quantity, multiplied by I_1. (or $I_1 X_p = \sqrt{E_z{}^2 - (I_1 R_p)^2}$) is the vector $E'_1 E_p$ of the diagrams in Figs. 35 and 36 as it might be determined experimentally for an actual transformer. If it were possible for all the magnetic flux to link with all the primary, and all the secondary, turns, the quantity $I_1 X_p$ would necessarily be zero; all the flux would be in the phase OB, and OE_z (of Fig. 38) would be equal to OP. The presence of the quantity $I_1 X_p$ can only be due to those flux lines which link with primary turns, but do not link with an equivalent number of secondary turns.

34. Calculation of Reactive Voltage Drop. Seeing that it is generally—although not always—desirable to obtain good regulation in transformers, it is obvious that designs with the primary and secondary windings on separate cores (see Figs. 1, 35 and 36), which greatly exaggerate the ratio of leakage flux to useful flux, would be very unsatisfactory in practice. By putting half the primary and half the secondary on each of the two limbs of a single-phase core-type transformer, as shown in Fig. 39, a considerable improvement is effected, but the reluctance of the leakage paths is still low, and this design is not nearly so good as Fig. 7 (page 18) where the leakage paths have a greater length in proportion to the cross-section. Similarly in the shell type of transformer, the design shown in Fig. 40 is unsatisfactory; the arrangement of

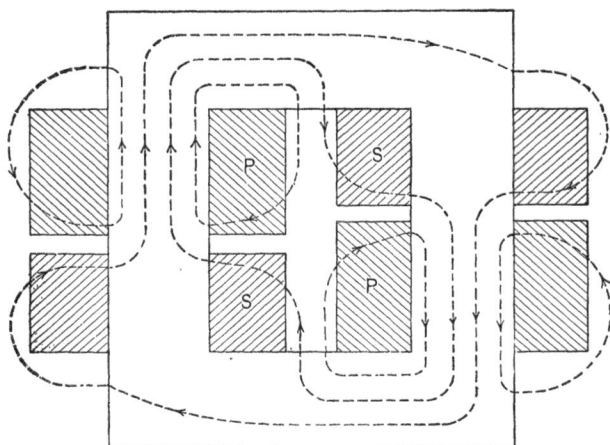

FIG. 39.—Leakage Flux Lines in Special Core-type Transformer.

coils, as shown in Figs. 10 and 11 (Art. 8) is much better because of the greater reluctance of the leakage paths. Transformers with coils arranged as in Figs. 7 and 10 are satisfactory for small sizes; but, in large units, it is necessary to subdivide the windings into a large number of sections with primary coils " sandwiched " between secondary coils as in Fig. 17 (core type) and Figs. 8 and

16 (shell type). By subdividing the windings in this manner, the m.m.f. producing the leakage flux, and the number of turns which this flux links with, are both greatly reduced.   The objection to a very large number of sections is the extra space taken up by insulation between the primary and secondary coils. For the purpose of facilitating calculations, the windings of transformers can generally be divided into unit sections

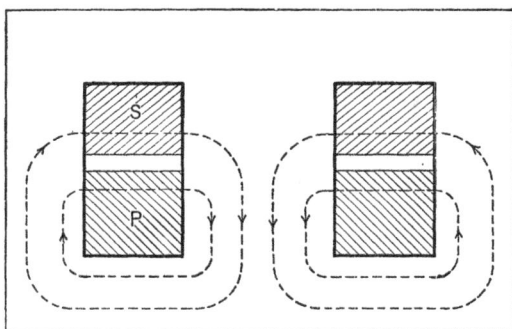

FIG. 40.—Leakage Flux Lines in Poorly Designed Shell-type Transformer.

as indicated in Fig. 41 (which shows an arrangement of coils in a shell-type transformer similar to Fig. 16). Each section consists of half a primary coil and half a secondary coil, with leakage flux passing through the coils and the insulation between them * all in the same

* If air ducts are required between sections of the winding, these should be provided in the position of the dotted center lines, by a further sub-division of each primary and secondary group of turns; thus allowing the space between primary and secondary coils to be filled with solid insulation. It is evident that, if good regulation is desired, the space between primary and secondary coils—where the leakage flux density has its maximum value—must be kept as small as possible.

direction, as indicated by the flux diagram at the bottom
of the figure.

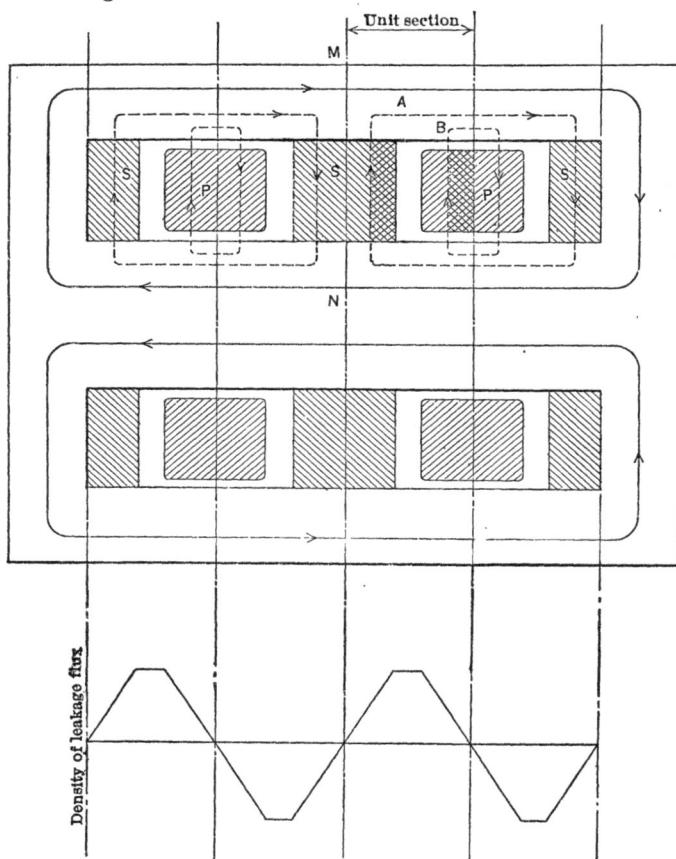

FIG. 41.—Section through Coils of Shell-type Transformer.

The effect of all leakage lines in the gap between the
coils is to produce a back e.m.f. in the primary without

affecting the voltage induced in the secondary by the main component of the total flux (represented by the full line). Of the other leakage lines, B links with only a portion of the primary turns and has no effect on the primary turns which it does not link with; while A links not only with all the primary turns, but also with a certain number of secondary turns. Note that if the line A were to coincide with the dotted center line MN, marking the limit of the unit section under consideration, it would have no effect on the transformer regulation because flux which links equally with primary and secondary is not leakage flux. Actually, the line A links with all the primary turns of the half coil in the section considered, but with only a portion of the secondary turns in the same section. Its effect is, therefore, exactly as if it linked with only a fractional number of the primary turns. The mathematical development which follows is based on these considerations.

Fig. 42 is an enlarged view of the unit section of Fig. 41, the length of which—measured perpendicularly to the cross section—is l cms. All the leakage is supposed to be along parallel lines perpendicular to the surface of the iron core above and below the coils.

It is desired to calculate the reactive voltage drop in a section of the winding of length l cms., depth h cms., and total width $(s+g+p)$ cms., where

s = the half thickness of the secondary coil;
g = the thickness of insulation between primary and secondary coils;
p = the half thickness of the primary coil.

The voltage drop caused by the leakage flux in the spaces g, p, and s will be calculated separately and then added together to obtain the total reactive voltage drop. The general formula giving the r.m.s. value of the volts induced by Φ maxwells linking with T turns is

$$IX = \frac{2\pi f}{10^8}\Phi T, \quad . \quad . \quad . \quad . \quad (36)$$

when the flux variation follows the simple harmonic law.

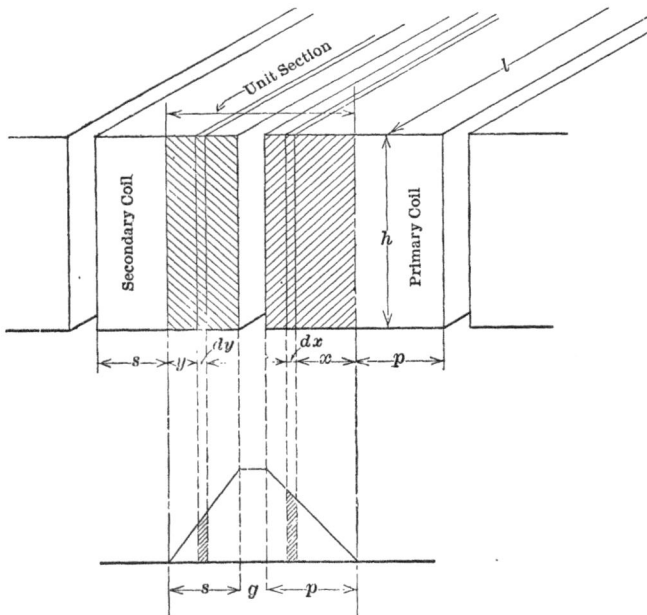

FIG. 42.—Enlarged Section through Transformer Coils.

In calculating the voltage produced by a portion of flux in a given path, we must therefore determine (1) the amount of this flux, and (2) the number of turns with which it links. The symbols T_1 and T_2 will be used to denote the number of turns in the half sections of widths p and s of the primary and secondary coils respectively. The meaning of the variables x and y is indicated in Fig. 42. The symbol m will be used for the quantity $\frac{2\pi f}{10^8}$.

For the section g we have,

$$(IX)_g = m\Phi T,$$

Inserting for Φ its value in terms of m.m.f. and permeance, this becomes,

$$(IX)_g = m(0.4\pi T_1 I_1) \times \frac{lg}{h} \times T_1. \quad . \quad . \quad (37)$$

In the section p, the m.m.f. producing the element of flux in the space of width dx is due to the current I_1 in $\left(\frac{x}{p}\right)T_1$ turns, and since this element of flux links with only $\left(\frac{x}{p}\right)T_1$ turns, we have,

$$d(IX)_p = m\left[0.4\pi\left(\frac{x}{p}\right)T_1 I_1\right]\frac{ldx}{h} \times \left(\frac{x}{p}\right)T_1,$$

whence

$$(IX)_p = m \times \frac{0.4\pi T_1^2 I_1 l}{p^2 h.} \int_0^p x^2 dx$$

$$= m \times 0.4\pi T_1^2 I_1 \frac{lp}{3h}. \quad . \quad . \quad . \quad . \quad (38)$$

In the section s, the m.m.f., producing the small element of flux in the space of width dy is due to the current I_s in $\left(\dfrac{y}{s}\right)T_s$ turns, and since this must be considered as linking with $\left(1-\dfrac{s-y}{s}\right)T_1$ turns, we can write,

$$d(IX)_s = m\left[0.4\pi\left(\frac{y}{s}\right)T_2 I_s\middle|\frac{ldy}{l}\left(\frac{y}{s}\right)T_1,\right.$$

whence

$$(IX)_s = m\times\frac{0.4\pi(T_2 I_s)T_1 l}{s^2 h}\int_0^s y^2 dy$$

$$= m\times 0.4\pi T_1^2 I_1\frac{ls}{3h}, \quad \cdots \quad (39)$$

wherefrom the secondary quantities T_2 and I_s have been eliminated by putting $(T_1 I_1)$ in place of $(T_2 I_s)$.

The final expression for the inductive voltage drop in the unit section considered is obtained by adding together the quantities (37), (38), and (39). Thus,

$$I_1 X_1 = \frac{2\pi f\times 0.4\pi T_1^2 I_1 l}{10^8 h}\left[g+\frac{p+s}{3}\right] \text{ volts}, \quad (40)$$

wherein all dimensions are expressed in centimeters.

If all the primary turns are connected in series, this quantity will have to be multipled by the ratio $\dfrac{T_p}{T_1}$ (or by twice the number of primary groups of coils) to obtain the value of the vector $I_1 X_p$ shown in the vector diagrams.

Equivalent Value of the Length l. The numerical value of the length l as used in the above formulas might reasonably be taken as the mean length per turn of the transformer windings, provided the reluctance of the flux paths outside the section shown in Fig. 42 may be neglected, not only where the iron laminations provide an easy path for the flux, but also where the ends of the coils project beyond the stampings.

Every manufacturer of transformers who has accumulated sufficient test data from transformers built to his particular designs, will be in a position to modify Formula (40) in order that it may accord very closely with the measured reactive voltage drop. This correction may be in the form of an expression for the equivalent length l, which takes into account the type of transformer (whether core or shell) and the arrangement of coils; or the quantity $\left[g+\dfrac{p+s}{3}\right]$ may be modified, being perhaps more nearly $\left[g+\dfrac{p+s}{2.5}\right]$, which allows for more leakage flux through the space occupied by the copper than is accounted for on the assumption of parallel flux lines. The writer believes, however, that if l is taken equal to the mean length per turn of the windings—expressed in centimeters—the Formula (40) will yield results sufficiently accurate for nearly all practical purposes.

35. Calculation of Exciting Current. Before drawing the complete transformer vector diagram, including the reactive drop calculated by means of the formula developed in the preceding article, it is necessary to consider how the magnitude and phase of the exciting current component of the total primary current may be pre-determined.

The exciting current (I_e) may be thought of as con-sisting of two components: (1) the magnetizing com-ponent (I_0) in phase with the main component of the magnetic flux, i.e., that which links with both primary and secondary coils, and (2) the " energy " component

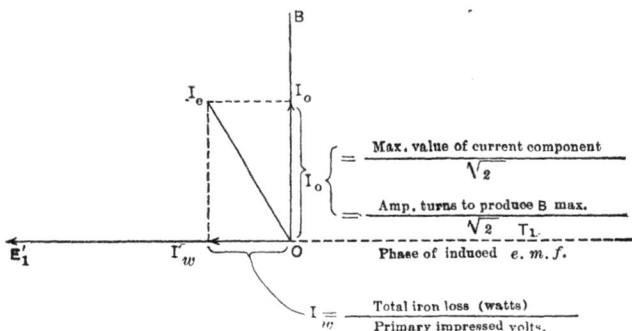

$$I_0 = \frac{\text{Max. value of current component}}{\sqrt{2}}$$

$$I_0 = \frac{\text{Amp. turns to produce B max.}}{\sqrt{2} \ T_1}$$

$$I_w = \frac{\text{Total iron loss (watts)}}{\text{Primary impressed volts.}}$$

FIG. 43.—Vector Diagram showing Components of Exciting Current.

(I_w) leading I_0 by one-quarter period, and, therefore, exactly opposite in phase to the induced e.m.f.

The magnitude of this component depends upon the amount of the iron losses only, because the very small copper losses $(I^2{}_e R_1)$ may be neglected.

If these components could be considered sine waves, the vector construction of Fig. 43 would give correctly the magnitude and phase of the total exciting current I_e. For values of flux density above the " knee " of the B-H curve, the instantaneous values of the magnetizing current are no longer proportional to the flux, and this component of the total exciting current cannot therefore be regarded as a sine wave even if the flux variations are sinusoidal. The error introduced by using the construction of Fig. 43 is, however, usually negligible because the exciting current is a very small fraction of the total primary current.

The notes on Fig. 43 are self explanatory, but reference should be made to Fig. 44 from which the ampere turns per inch of the iron core may be read for any value of the (maximum) flux density. The flux density is given in gausses, or maxwells per square centimeter of cross-section.* The total magnetizing ampere turns are equal to the number read off the curve multiplied by the mean length of path of the flux which links with both primary and secondary coils. When butt joints are present in the core, the added reluctance should be allowed for. Each butt joint may be considered as an air gap 0.005 in. long, and the ampere turns to be allowed in addition to those for the iron portion of the magnetic circuit are therefore,

$$Amp. \ turns \ for \ joints = \frac{Hl}{0.4\pi}$$

$$= 0.01 \times B_{max} \times \text{No. of butt joints in series.} \quad (41)$$

Instead of calculating the exciting current by the method outlined above, designers sometimes make use

---

* The writer makes no apology for using both the inch and the centimeter as units of length. So long as engineers insist that the inch has certain inherent virtues which the centimeter does not possess, they should submit without protest to the inconvenience and possible disadvantage of having to use conversion factors, especially in connection with work based on the fundamental laws of physics.

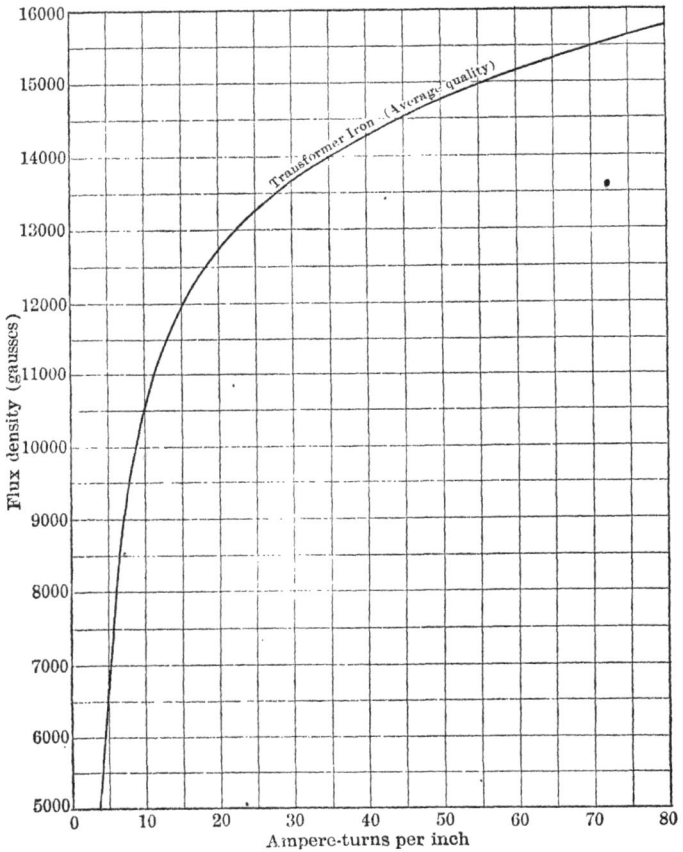

FIG. 44.—Curve giving Connection between Magnetizing Ampere-turns and Flux Density in Transformer Iron.

of curves connecting maximum core density and volt-amperes of total exciting current per cubic inch or per pound of core; the data being obtained from tests on completed transformers. The fact that the total volt-amperes of excitation (neglecting air gaps) are some function of the flux density multiplied by the weight of the iron in the transformer core, may be explained as follows:

Let w = total watts lost per pound of iron, corresponding to a particular value of B as read off one of the curves of Fig. 27;

a = Ampere turns per inch as read off Fig. 44;

A = cross-section of iron in the core, measured perpendicularly to the magnetic flux lines (square inches);

l = Length of the core in the direction of the flux lines (inches);

P = Weight of core in pounds = $0.28Al$.

The symbols previously used are:

T_p = number of primary turns:

$$E_p = \text{primary volts} = \frac{4.44T_p(6.45BA)f}{10^8}.$$

Given definite values for B and f, the "in phase" component of the exciting current is

$$I_w = \frac{\text{core loss}}{E_p} = \frac{w \times P}{E_p},$$

and the " wattless " component, or true magnetizing current, is

$$I_0 = \frac{a \times l}{T_p},$$

whence

$$I_e = \sqrt{\left(\frac{wP}{E_p}\right)^2 + \left(\frac{al}{T_p}\right)^2}.$$

Multiplying both sides of the equation by $\frac{E_p}{P}$, we get

$$\frac{E_p I_e}{P} = \frac{\text{volt-amperes of total excitation}}{\text{weight of core}}$$

$$= \sqrt{w^2 + \left(\frac{4.44a6.45Bf}{0.28 \times 10^8}\right)^2}. \quad \cdots \quad (42)$$

This formula may be used for plotting curves such as those in Fig. 45. Thus, if

$B = 13{,}000$ gausses,
$f = 60$ cycles per second,
$w = 1.55$ (read off curve for silicon steel in Fig. 27),
$a = 22$ (from Fig. 44); and, by Formula (42)

Volt-amperes per pound

$$= \sqrt{(1.55)^2 + \left(\frac{4.44 \times 22 \times 13{,}000 \times 6.45 \times 60}{0.28 \times 10^8}\right)^2} = 17.8.$$

The error in this method of deriving the curves of Fig. 45 is due to the fact that sine waves are assumed. The data for plotting the curves should properly be obtained from tests on cores made out of the material to be used in the construction of the transformer.

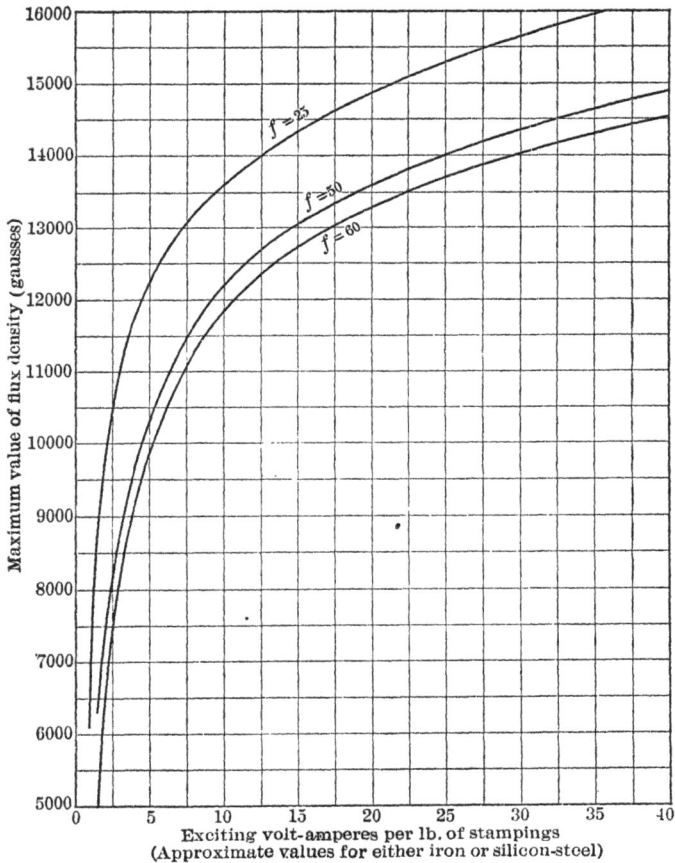

FIG. 45.—Curves giving Connection between Exciting Volt-amperes and Flux Density in Transformer Stampings.

The effect of the magnetizing current component in distorting the current waves may be appreciable when the core density is carried up to high values. The curve of flux variation cannot then be a sine wave, and the introduction of high harmonics in the current wave may aggravate the disturbances that are always liable to occur in telephone circuits paralleling overhead transmission lines. This is one reason why high values of the exciting current are objectionable. An open-circuit primary current exceeding 10 per cent of the full-load current would rarely be permissible.

36. Vector Diagrams Showing Effect of Magnetic Leakage on Voltage Regulation of Transformers. The vector diagrams, Figs. 46, 47, and 48, have been drawn to show the voltage relations in transformers having appreciable magnetic leakage. The proportionate length of the vectors representing IR drop, IX drop, and magnetizing current, has purposely been exaggerated in order that the construction of the diagrams may be easily followed.

Fig. 46 is the complete vector diagram of a transformer; the meaning of the various component quantities being as follows:

$E_2 =$ Induced secondary e.m.f., due to the flux (OB) linking with the secondary turns;

$E_s =$ Secondary terminal voltage when the secondary current is I_s amperes on a load power factor of cos θ;

$I_e =$ Primary exciting current, calculated as explained in the preceding article;

$I_1 =$ Balancing component of total primary current

$$\left(= I_s \times \frac{T_s}{T_p} \right);$$

$I_p =$ Total primary current;

$E'_1 =$ Balancing component of induced primary volt-

age $\left(= E_2 \times \frac{T_p}{T_s} \right);$

$PE'_1 = IR$ drop due to primary resistance (drawn parallel to OI_p);

$E_pP = IX$ drop due to leakage reactance (drawn at right angles to OI_p);

$E_p =$ Impressed primary e.m.f.

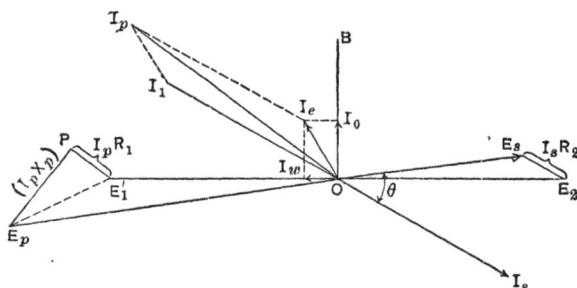

FIG. 46.—Vector Diagram of Transformer on Inductive Load.

It is usually permissible to neglect the exciting current component when considering full-load conditions. This leads to the simpler diagram, Fig. 47, in which the total primary current is supposed to be of the same magnitude and phase as what has previously been referred to as the balancing component of the total primary current.

The dotted lines in Fig. 47 show how a still greater simplification may be effected in drawing a vector diagram from which the voltage regulation can be calculated. Instead of drawing the two vectors OE_2 and OE_s for the induced and terminal secondary voltages, we can draw OE_t opposite in phase to E_s and equal to $E_s\left(\dfrac{T_p}{T_s}\right)$. Then E_eP (drawn parallel to OI_1) is the component of the impressed primary volts necessary to overcome the ohmic resistance of both primary and secondary windings.

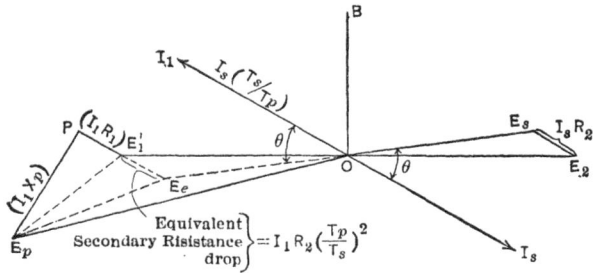

FIG. 47.—Simplified Vector Diagram of Transformer; Exciting Current Neglected.

It is now only necessary to turn this diagram through 180 degrees, and eliminate all unnecessary vectors, in order to arrive at the very simple diagram of Fig. 48, from which the voltage regulation can be calculated.

37. Formulas for Voltage Regulation. From an inspection of Fig. 48, it is seen that

$$E_p = \frac{(I_1R_p) + E_e \cos\theta}{\cos\phi}, \quad \cdot \quad \cdot \quad \cdot \quad (43)$$

wherein $\cos \theta$ is known (being the power factor of the external load), and $\cos \phi$ has not yet been determined. But,

$$\tan \phi = \frac{(I_1 X_p) + E_e \sin \theta}{(I_1 R_p) + E_e \cos \theta}, \quad \cdot \quad \cdot \quad \cdot \quad (44)$$

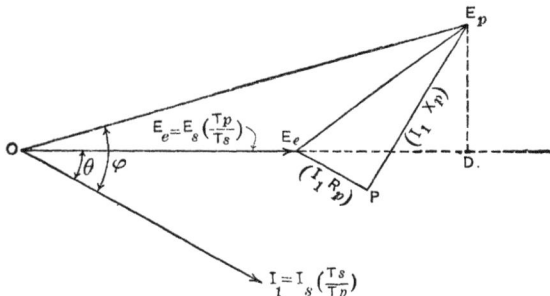

Fig. 48.—Simple Transformer Vector Diagram for Calculation of Voltage Regulation.

which can be used to calculate ϕ and therefore $\cos \phi$.

The percentage regulation is

$$100 \times \frac{E_p - E_e}{E_e} = 100 \times \frac{I_1 R_p + E_e (\cos \theta - \cos \phi)}{E_e \cos \phi}, \quad (45)$$

or, if the ohmic drop is expressed as a percentage of the (lower) terminal voltage:

Per cent regulation

$$= \frac{\text{Per cent equiv. } IR \text{ drop} + 100(\cos \theta - \cos \phi)}{\cos \phi}. \quad (46)$$

The difference between the angles θ and ϕ (Fig. 48) is generally small, and it is then permissible to assume that $OD = OE_p$. But

$$OD = E_e + I_1 R_v \cos \theta + I_1 X_p \sin \theta,$$

whence,

Per cent regulation (approximate)

$$= \text{Per cent } IR \cos \theta + \text{per cent } IX \sin \theta. \quad (47)$$

If the power factor were leading instead of lagging as in Fig. 48, the *plus* sign would have to be changed to a *minus* sign.

Example. In order to show that the approximate Formula (47) is sufficiently accurate for practical purposes, the following numerical values are assumed.

Power factor $(\cos \theta) = 0.8$.
Total IR drop $= 1.5$ per cent.
Total IX drop $= 6.0$ per cent.

By Formula (44),

$$\tan \phi = \frac{0.06 + 0.6}{0.015 + 0.8} = 0.81,$$

whence $\cos \phi = 0.777$, and, by Formula (46),

$$\text{Regulation} = \frac{1.5 + 80 - 77.7}{0.777} = 4.9 \text{ per cent.}$$

By the approximate Formula (47),

Regulation $= (1.5 \times 0.8) + (6 \times 0.6) = 4.8$ per cent.

The total equivalent voltage drop, due to the resistance of the windings (the quantity $I_1 R_p$ of the vector diagrams) is usually between 1 and 2 per cent of the terminal voltage in modern transformers. The reactive voltage drop caused by magnetic leakage (the quantity $I_1 X_p$ in the vector diagrams) is nearly always greater than the IR drop, being 3 to 8 per cent of the terminal voltage. Sometimes it is 10 per cent, or even more, especially in high-voltage transformers where the space occupied by insulation is considerable, or in transformers of very large size, when the object is to keep the current on short circuit within safe limits.

CHAPTER V

37. The Output Equation. The volt-ampere output of a single-phase transformer is $E \times I$ which, as explained in Art. 6, may be written

$$\text{Volt-amperes} = \frac{4.44f}{10^8} \times \Phi \times (TI), \quad . \quad . \quad (48)$$

where TI stands for the total ampere turns of either the primary or secondary winding.

There is no limit to the number of designs which will satisfy this equation; the total flux, Φ, is roughly a measure of the cross-section of the iron core, while the quantity (TI) determines the cross-section of the windings. The problem before the designer is to proportion the parts and dispose the material in such a way as to obtain the desired output and specified efficiency at the lowest cost. The temperature rise is also a matter of importance which must be watched, and light weight is occasionally more important than cost.

It cannot be said that there is one method of attacking the problems of transformer design which has indisputable advantages over all others; and in this, as in all design, the judgment and experience of the individual designer must necessarily play an important part. The apparent simplicity of the calculations involved in transformer design is the probable cause of the many more or less unsuccessful attempts to reach the desired end by purely mathematical methods. It is not possible to include all the variable factors in practical mathematical equations purporting to give the ideal quantities and proportions to satisfy the specification. Methods of procedure aiming to dispense with individual judgment and a certain amount of correction or adjustment in the final design, should generally be discountenanced, because they are based on inadequate or incorrect assumptions which are liable to be overlooked as the work proceeds and becomes finally crystallized into more or less formidable equations and formulas of unwieldy proportions.

No claim to originality is made in connection with the following method of procedure; indeed it is questionable whether the mass of existing literature treating of the alternating current transformer leaves anything new to be said on the subject of procedure in design. All that the present writer hopes to present is a treatment consistent with what has gone before, based always on the fundamental principles of physics—even though the use of empirical constants may be necessary.

Instead of attempting to take account at one time of all the conditions to be satisfied in the final design, the factors which have the greatest influence on the dimensions will be considered first; items such as temperature rise and voltage regulation being checked later and, if necessary, corrected by slight changes in the dimensions or proportions of the preliminary design.

38. Specifications. It will be advisable to list here the particulars usually specified by the buyer, and supplement these, if required, with certain assumptions that the manufacturer must make before he can proceed with a particular design.

(1) K.v.a. output.

(2) Number of phases.

(3) Primary and secondary voltages (E_p and E_s).

(4) Frequency (f).

(5) Efficiency under specified conditions.

(6) Voltage regulation under specified load.

(7) Method of cooling—Temperature rise.

(8) Maximum permissible open-circuit exciting current.

Items (1) to (4) must always be stated by the purchaser, while the other items may be determined by the manufacturer, who should, however, be called upon to furnish these particulars in connection with any competitive offer.

With reference to item (5), if the efficiency is stated for two different loads, the permissible copper and iron losses can be calculated. If the buyer does not furnish these particulars, he should state whether the transformer is for use in power stations or on distributing lines, in order that the relation of the iron losses to the total losses may be adjusted to give a reasonable all-day efficiency. In any case, before proceeding with the design, the maximum permissible iron and copper losses must be known or assumed.

The requirements of items (6), (7), and (8), are to some

extent satisfied, even in the preliminary design, by selecting a flux density (B) and a current density (Δ) from the values given in Article 20, because industrial competition and experience have shown these values to·give the best results while using the smallest permissible amount of material. Thus, by selecting a proper value for Δ, both the local heating and the IR drop of the windings will probably be within reasonable limits. The other factor influencing the voltage regulation (item (6)) is the reactive drop, which can generally be controlled by suitably subdividing the windings.

A proper value of the flux density (B) will generally keep item (8) within the customary limits.

39. Estimate of Number of Turns in Windings. Returning to the Formula (48) in Article 37, if a suitable value for T could be determined or assumed, the only unknown quantity in the output equation would be Φ and we should then have a starting-point from which the dimensions of a preliminary design could be easily calculated.

Let V = volts per turn (of either primary or secondary winding) then, in order to express this quantity in terms of the volt-ampere output, we have,

$$V_t = \frac{E}{T} = \frac{(EI)}{TI},$$

from which T must be eliminated, since the reason for seeking a value for V_t is that T may be calculated therefrom.

Using the value of (EI) as given by Formula (48), we can write

$$V_t^2 = \frac{EI}{TI} \times \frac{4.44 f \Phi TI}{TI \times 10^8},$$

whence

$$V_t = \sqrt{\text{volt-ampere output}} \times \sqrt{\frac{4.44}{10^8}\left(\frac{f\Phi}{TI}\right)}. \qquad (49)$$

The quantity in brackets under the second radical is found to have an approximately constant value, for an efficient and economical design of a given type, without reference to the output. This permits of the formula being put in the form

$$V_t = c \times \sqrt{\text{volt-ampere output}}, \qquad \ldots \quad (49a)$$

where c is an empirical coefficient based on data taken from practical designs.

Factors Influencing the Value of the Coefficient c.
It is proposed to examine the meaning of the ratio $\dfrac{f\Phi}{TI}$ which appears under the second radical of Formula (49) with a view to expressing this in terms of known quantities, or of quantities that can easily be estimated.

Let W_c = full load copper losses (watts);

W_i = core losses (watts);

the relation between these losses being;

$$W_c = bW_i, \qquad \ldots \quad \ldots \quad (50)$$

wherein b must always be known before proceeding with the design.

Let l_c = mean length per turn of copper in windings;

l_i = mean length of magnetic circuit measured along flux lines;

then

$$W_c = constant \times \Delta^2 \times volume\ of\ copper$$

$$= k_c(TI)\Delta l_c. \qquad \ldots \ldots \ldots \quad (51)$$

where k_c is a constant to be determined later.   Similarly

$$W_i = constant \times fB^n \times volume\ of\ iron$$

$$= k_i fB^n \left(\frac{\Phi}{B}\right) l_i$$

$$= k_i(f\Phi)B^{n-1} l_i, \qquad \ldots \ldots \ldots \quad (52)$$

wherein k_i is another constant to be determined later.

Inserting these values in Formula (50), the required ratio can be put in the form

$$\frac{f\Phi}{TI} = \frac{k_c\Delta}{bk_iB^{n-1}} \left(\frac{l_c}{l_i}\right). \qquad \ldots \ldots \ldots \quad (53)$$

This ratio is thus seen to depend on certain quantities and constants which are only slightly influenced by the *output* of the transformer.   They depend on such items as the ratio of copper losses to iron losses (i.e., whether the transformer is for use on power transmission lines, or distributing circuits); temperature rise and methods of cooling; space factor (voltage); and also on the type—whether core or shell—since this affects the best relation between mean lengths of the copper and iron circuits.

The Factor k_c. Using the inch for the unit of length, and allowing 7 per cent for eddy-current losses in the copper, the resistivity of the windings will be 0.9×10^{-6} ohms per inch-cube at a temperature of 80° C.; the loss per cubic inch of copper $= \Delta^2 \times 0.9 \times 10^{-6}$, and since the volume is $2\left(\dfrac{TI}{\Delta}\right)l_c$, it follows that $k_c = 2 \times 0.9 \times 10^{-6}$.

The Factor k_i. If $w = $ total watts lost per pound as read off one of the curves of Fig. 27, and if l_i is in inches, we have the equation

$$0.28w\left(\frac{\Phi}{6.45B}\right)l_i = k f \Phi B^{n-1} l_i,$$

whence

$$k_i = \frac{0.28w}{6.45 f B^n}.$$

The Factor b. The ratio of full-load copper loss to iron loss will determine the load at which maximum efficiency occurs.

Let us assume the k.v.a. output and the frequency of a given transformer to be constant, and determine the conditions under which the total losses will be a minimum. It is understood that, if the current I is increased, the voltage, E, must be decreased; but the condition k.v.a. $= EI$ must always be satisfied.

The sum of the losses is $W_c + W_i$; but

$$W_c = I^2 R = \frac{(\text{k.v.a.})^2}{E^2} R = \frac{\text{a constant}}{E^2},$$

and

$$W_i \propto B^n.$$

Also, since f remains constant, $E \propto B$, and we can write

$$W_i = \text{a constant} \times E^n.$$

The quantity which must be a minimum is therefore

$$\frac{\text{a constant}}{E^2} + \text{a constant} \times E^n.$$

If we take the differential coefficient of this function of E and put it equal to zero, we get the relation

$$\frac{W_c}{W_i} = \frac{n}{2}.$$

The value of n for high densities is about 2, while for low densities it is nearer to 1.7, a good average being 1.85. Thus, to obtain maximum efficiency at full load in a power transformer, the ratio of copper loss to iron loss should be about $b = \frac{1.85}{2} = 0.925$.

In a distributing transformer, in order to obtain a good all-day efficiency, the maximum efficiency should occur at about $\frac{2}{3}$ full load, whence

$$\frac{(\frac{2}{3})^2 W_c}{W_i} = \frac{n}{2}.$$

Taking $n = 1.75$, because of the lower densities generally used in small self-cooling transformers, we get

$$b = \frac{1.75 \times 9}{4 \times 2} = 1.97 \quad \text{or (say)} \quad 2.$$

The Ratio $\frac{l_c}{l_i}$. Considerable variations in this ratio are permissible, even in transformers of a given type wound for a particular voltage, and that is one reason why a close estimate of the volts per turn as given by Formula (49) is not necessary. Refinements in proportioning the dimensions of a transformer are rarely justified by any appreciable improvement in cost or efficiency; a certain minimum quantity of material is required in order to keep the losses within the specified limits; but considerable changes in the shape of the magnetic and electric circuits can be made without greatly altering the total cost of iron and copper, provided always that the important items of temperature rise and regulation are checked and maintained within the specified limits.

Figs. 49 and 50 show the assembled iron stampings of single-phase shell- and core-type transformers. The proportions will depend somewhat upon the voltage and method of cooling; but if the leading dimensions are expressed in terms of the width (L) of the stampings under the coils, they will generally be within the following limits:

Shell Type.	Core Type.
$S = 2$ to 3 times L	$S = 1$ to 1.8 times L
$B = 0.5$ to 0.75 times L	$B = 1$ to 1.5 times L
$D = 0.6$ to 1.2 times L	$D = 1$ to 2 times L
$H = 1.2$ to 3.5 times L	$H = 3$ to 6 times L

By taking the averages of these figures, and roughly approximating the lengths l_c and l_i in each case, the mean value of the required ratio is found to be

$$\frac{l_c}{l_i} = 1.2 \text{ (approx.) for shell type,}$$

$$\frac{l_c}{l_i} = 0.3 \text{ (approx.) for core type.}$$

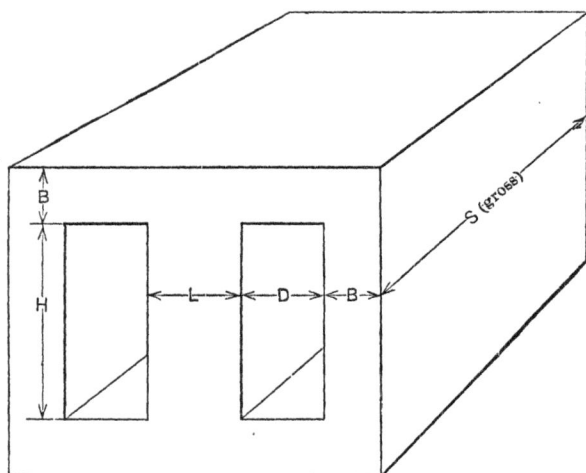

FIG. 49.—Assembled Stampings of Single-phase Shell-type Transformer.

Having determined the values of the various quantities appearing in Formula (53), it is now possible to calculate an approximate average value for the quantity $\frac{f\Phi}{TI}$ and for the coefficient c of Formula (49).

We shall make the further assumptions (refer Art. 20) that $\Delta = 1100$ amperes per square inch, and $B = 8000$

gausses; the transformer being of the shell type for use on distributing circuits of frequency 60. Then, by Formula (53),

$$\frac{f\Phi}{Tl} = \frac{2 \times 0.9 \times 1100 \times 1.2 \times 6.45 \times 60 \times 9000}{10^6 \times 2 \times 0.28 \times 0.75} = 19,720$$

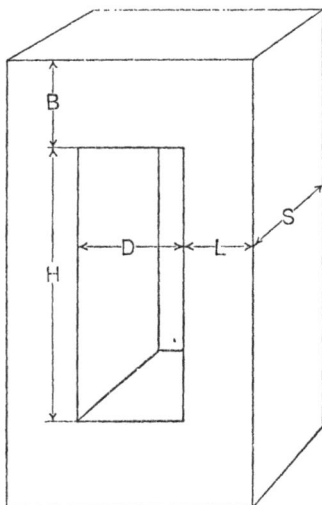

FIG. 50.—Assembled Stampings of Single-phase Core-type Transformer.

wherein the figure 0.75 is the value of w read off the curve for silicon steel in Fig. 27.

The value of the coefficient in Formula (49), for the assumed conditions, is therefore

$$c = \sqrt{\frac{4.44}{10^8} \times 19,720} = 0.0296.$$

Similarly, for a core-type power transformer; if $f = 25$. $B = 13,000$, and $\Delta = 1350$, we have,

$$\frac{f\Phi}{TI} = \frac{2 \times 0.9 \times 1350 \times 0.5 \times 6.45 \times 25 \times 13,000}{10^6 \times 0.925 \times 0.28 \times 0.58} = 16,940.$$

Whence $c = 0.0274$.

Having shown what factors determine this design coefficient, it will merely be necessary to give a list of values from which a selection should be made for the purpose of calculating the quantity V_t of Formula $(49a)$.

For shell-type power transformers $c = 0.04$ to 0.045
For shell-type distributing transformers $c = 0.03$
For core-type power transformers $c = 0.025$ to 0.03
For core-type distributing transformers $c = 0.02$

Where a choice of two values of c is given, the lower value should be chosen for transformers wound for high pressures. When the voltage is low the value of c is slightly higher because of the alteration in the ratio. $\frac{l_c}{l_i}$ which depends somewhat on the copper space factor. The proposed values here given for this design coefficient are based on the assumption that silicon steel stampings are used in the core. If ordinary transformer iron is used—as, for instance, in small distributing transformers—it will be advisable to take about $\frac{3}{4}$ of the above values for the coefficient c.

40. Procedure to Determine Dimensions of a New Design. With the aid of the design coefficient c, it is now possible to calculate the number of volts that should be generated in one turn of the winding of a transformer of good design according to present knowledge and practice. The logical sequence of the succeeding steps in the design, may be outlined as follows:

(1) Determine approximate dimensions.

    (a) Calculate volts per turn by Formula (49).

    (b) Assume current density (select suitable trial value from table in Art. 20). Decide on number of coils. Calculate cross-section of copper.

    (c) Decide upon necessary insulation and oil- or air-ducts between coils, and between windings and core. Determine shape and size of " window " or opening necessary to accommodate the windings.

    (d) Calculate total flux required. Assume flux density (select suitable trial value from table in Art. 20), and calculate cross-section of core. Decide upon shape and size of section, including oil- or air-ducts if necessary.

    (e) Calculate iron and copper losses, and modify the design slightly if necessary to keep these within the specified limits.

(2) Calculate approximate weight and cost of iron and copper if desired to check with permissible maximum before proceeding with the design.

(3) Calculate exciting current.

(4) Calculate leakage reactance and voltage regulation.

(5) Calculate necessary cooling surfaces. Design containing tank and lid, providing not only sufficient oil capacity and cooling surface, but also the necessary clearances to insure proper insulation between current-carrying parts and the case. Calculate temperature rise.

41. Space Factors. The copper space factor, as previously defined (see Art. 15), is the ratio between the cross-section of copper and the area of the opening or " window " which is necessary to accommodate this copper together with the insulation and oil- or air-ducts. It may vary between 0.55 in transformers for use on circuits not exceeding 660 volts, to 0.06 in power transformers wound for about 100,000 volts. An estimated value of the probable copper space factor may be useful to the designer when deciding upon one of the dimensions of the " window " in the iron core. For this purpose, the curves of Fig. 51 may be used, although the best design and arrangement of coils and ducts will not always lead to a space factor falling within the limits included between these two curves.

Iron Space Factor. The so-called stacking factor for the iron core will be between 0.86 and 0.9, and the total thickness of core, multiplied by this factor, will give the net thickness of iron if there are no oil- or air-ducts. When spaces are left between sections of the core for air or oil circulation, the iron space factor may be from 0.65 to 0.75.

42. Weight and Cost of Transformers. The weight per k.v.a. of transformer output depends not only upon the total output, but also upon the voltage and frequency. The net and gross weights of particular trans-

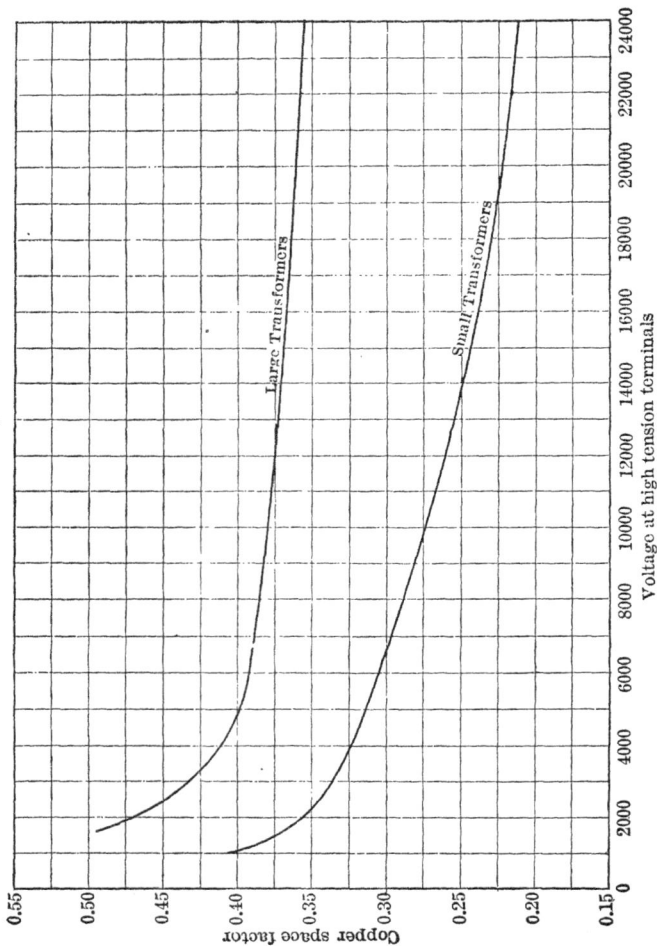

FIG. 51.—Copper Space Factors.

formers can be obtained from manufacturers' catalogues
and also from the Handbooks for Electrical Engineers.
The effect of output and frequency on the weight of a
line of transformers designed for a particular voltage
(in this instance, 22,000 volts) is roughly indicated by
the following figures of weight per k.v.a. of output.
These figures include the weight of oil and case.

$$\text{Frequency 60} \begin{cases} \text{100 k.v.a. output} & \text{40 lb.} \\ \text{500 k.v.a. output} & \text{23 lb.} \end{cases}$$

$$\text{Frequency 25} \begin{cases} \text{100 k.v.a. output} & \text{52 lb.} \\ \text{500 k.v.a. output} & \text{35 lb.} \end{cases}$$

The cost of transformers, depending as it does on the
fluctuating prices of copper and iron, is very unstable.
Within the last few years, the variation in the price of
copper wire has been about 100 per cent, and the cost of
the laminated iron for the cores has also undergone great
changes. The best that can be done here is to indicate
how the cost depends upon voltage and output.   That a
high frequency always means a cheaper transformer is
evident from an inspection of the fundamental Formula
(48) of Art. 37. If f is increased, either Φ, or (TI), or
both, can be reduced, and this means a saving of iron,
or copper, or both.   The effect of an increase in voltage
is felt particularly in the smaller sizes, but an increase
of voltage always means an addition to the cost; while
an increase of size for a given voltage results in a reduction of the cost per k.v.a. of output.
Some idea of the dependence of cost on output and
voltage may be gained from the fact that the unit cost

would be about the same for (1) a 1500 k.v.a. transformer wound for 22,000 volts, (2) a 2000 k.v.a. transformer wound for 44,000 volts, and (3) a 3000 k.v.a. transformer wound for 88,000 volts.

Three-phase Transformers. It does not appear to be necessary to supplement what has been said in Articles 5 and 8 on the subject of three-phase transformers. Once the principles underlying the design of single-phase transformers are thoroughly understood, it is merely necessary to divide any polyphase transformer (see Figs. 12, 13, and 14) into sections which can be treated as single-phase transformers, due attention being paid to the voltage and k.v.a. capacity of each such unit section of the three-phase transformer.

The saving of materials effected by combining the magnetic circuits of three single-phase transformers so as to produce one three-phase unit, usually results in a reduction of 10 per cent in the weight and cost.

43. Numerical Example. It is proposed to design a single-phase 1500 k.v.a. oil-insulated, water-cooled, transformer for use on an 88,000-volt power transmission system. A design sheet containing more detailed items than would generally be considered necessary will be used in order to illustrate the various steps in the design as developed and discussed in the preceding articles. Two columns will be provided for recording the known or calculated quantities, the first being used for preliminary assumptions or tentative values, while the second will be used for final results after the preliminary values have been either confirmed or modified.

SPECIFICATION

Output............................. 1,500 k.v.a.

Number of phases.................... one

H.T. voltage....................... 88,000

L.T. voltage....................... 6,000

Frequency.......................... 50

Maximum efficiency, to occur at full load
 and not to be less than............. 98.1%

Voltage regulation, on 80 per cent power
 factor........................... 5%

Temperature rise after continuous full-
 load run.......................... 40° C.

Test voltage: H.T. winding to case and
 L.T. coils........................ 177,000

L.T. winding to case................ 14,000

The calculated values of the various items are here brought together for reference and for convenience in following the successive steps in the design. The items are numbered to facilitate reference to the notes and more detailed calculations which follow.

Items (1) *and* (2). - *L.T. Winding.* By Formula (49a), Art. 39, page 142, the volts per turn, for a shell-type power transformer, are

$$V_t = 0.042 \sqrt{1,500,000} = 51.5,$$

whence,

$$T_s = \frac{6600}{51.5} = 128.$$

DESIGN SHEET

	Symbol.	Assumed or Approximate Values.	Final Values.
1. Volts per turn...................	V_t	51.5	52.3
L.T. WINDING (SECONDARY)			
2. Total number of turns.............	T_s	128	126
3. Number of coils...................	.....	6	6
4. Number of turns per coil...........	.....	21.3	21
5. Secondary current, amperes........	I_s	......	227
6. Current density, amperes per sq. in...	. Δ	1600	1575
7. Cross-section of each conductor, sq. in.	3 strips, each 0.16×0.3=0.144		
8. Insulation on wire, cotton tape, in...	.....	....	0.026
9. Insulation between layers, in........	(2×0.006)+0.012=0.024		
10. Number of turns per layer, per coil...	.....	....	1
11. Number of layers..................	.....	....	21
12. Overall width of finished coil (say), in.	.....	....	0.36
13. Thickness (or depth) of coil, with allowance for irregularities and bulging at center, in..............	.....	....	11.5
H.T. WINDING (PRIMARY)			
14. Total number of turns..............	T_p	....	1680
15. Number of coils...................	.....	18	18
16. Number of turns per coil...........	80 in 2 coils; 95 in 16 coils		
17. Primary current, amperes..........	I_p	17.05	
18. Current density, amperes per sq. in..	Δ	1640	
19. Cross-section of each wire, sq. in....	0.04×0.26=0.0104		
20. Insul. on wire (cotton covering), in...	2×0.008=0.016		
21. Insul. between layers, fullerboard, in..	.....	....	0.012
22. Number of turns per layer, per coil...	.....	....	1
23. Number of layers; in all but end coils	.....	....	95
24. Overall width of finished coil, in.....	.....	....	0.31
25. Thickness or depth of coil, in........	.....	....	6.75
26. Make sketch of assembly of coils, with necessary insulating spaces and oil ducts.			

DESIGN SHEET—*Continued*

	Symbol.	Assumed or Approximate Values.	Final Values.
27. Size of "window" or opening for windings, in.	.....	12.75×32	
MAGNETIC CIRCUIT			
28. Total flux (maxwells)...............	Φ	2.36×10^7	
29. Maximum value of flux density in core under windings (gausses).....	B	13,000	13,850
30. Cross-section of iron under coils, sq. in.	.....	282	264
31. Number of oil ducts in core.........	.....	none	
32. Width of oil ducts in core..........			
33. Width of stampings under windings, in...............................	L	11	11
34. Net length of iron in core, in.........	.....	....	24
35. Gross length of core, in.............	S	....	27
36. Cross-section of iron in magnetic circuit outside windings, sq. in.......	.....	....	264 ·
37. Flux density in core outside windings (gausses)......................	B	....	13,850
38. Average length of magnetic circuit under coils, in....................	.....	....	111.5
39. Average length of magnetic circuit outside coils, in..................	.....	....	111.5
40. Weight of core, lb.	.....	....	8250
41. Losses in the iron, watts............	.....	W_t	11,900
COPPER LOSSES			
42. Mean length per turn of primary, ft..	.....	....	10.15
43. Resistance of primary winding, ohms.	R_1	....	18.1
44. Full-load losses in primary (exc. current neglected)...................	.....	....	5360
45. Mean length per turn of secondary, ft.	.....	....	10.15
46. Resistance of secondary winding, ohms	R_2	....	0.0962
47. Full-load losses in secondary, watts ..	.....	....	4960
48. Total full-load copper losses, watts..	.....	W_c	10,220

DESIGN SHEET—*Continued*

	Symbol.	Assumed or Approximate Values.	Final Values.
49. Total weight of copper in windings, lb.	.....	....	1700
50. Efficiency at full load (unity power factor)............................	.....	....	0.985
51. Efficiency at other loads and power factors (refer to text following).			
52. No-load primary exciting current, amperes........................	I_e	....	2.15
REGULATION			
53. Reactive voltage drop..............	I_1X_p	....	2850
54. Equivalent ohmic voltage drop......	I_1R_p	....	600
55. Regulation on unity power factor, per cent............................	.....	....	0.735
56. Regulation on 80 per cent power factor, per cent.....................	.....	....	2.5
DESIGN OF TANK, COOLING SURFACES			
57. Effective cooling surface of tank, sq. in.	.....	19,360	18,860
58. Number of watts dissipated from tank surface........................	.....	4650	
59. Watts to be carried away by circulating water......................	.....	17,470	.
60. Size and length of pipe in cooling coil,.	.....	$1\frac{1}{4}''\times370'$	
61. Approximate flow of water per minute, gallons........................	.....	4.37	
62. Approximate weight of oil, lb........	.....	....	7300
63. Estimated total weight of transformer, lb.............................	.....	....	22,000

Items (3) *and* (4). The number of separate coils is determined by the following considerations:

(*a*) The voltage per coil should preferably not exceed 5000 volts.

(b) The thickness per coil should be small (usually within 1.5 in.) in order that the heat may readily be carried away by the oil or air in the ducts between coils (Refer Art. 23).

(c) The number of coils must be large enough to admit of proper subdivision into sections of adjacent primary and secondary coils to satisfy the requirements of regulation by limiting the magnetic flux-linkages of the leakage field.

(d) An even number of L.T. coils is desirable in order to provide for a low-tension coil near the iron at each end of the stack.

To satisfy (a), there must be at least $\dfrac{88,000}{5000}$ or, say 18 H.T. coils. If an equal number of secondary coils were provided, we could, if desired, have as many as eighteen similar high-low sections which would be more than necessary to satisfy (c). The number of these high-low sections or groupings must be estimated now in order that the arrangement of the coils, and the number of secondary coils, may be decided upon with a view to calculating the size of the " windows " in the magnetic circuit. It is true that the calculations of reactive drop and regulation can only be made later; but these will check the correctness of the assumptions now made, and the coil grouping will have to be changed if necessary after the preliminary design has been carried somewhat farther. The least space occupied by the insulation, and the shortest magnetic circuit, would be obtained by grouping all the primary coils in the center, with half the secondary winding at each end, thus giving only two high-low sections; but this would lead to a very high leakage reactance, and regulation much worse than the specified 6 per cent. Experience suggests that about six high-low sections should suffice in a transformer of this size and voltage, and we shall try this by arranging the high-tension coils in groups of six, and providing six secondary coils (see Fig. 52). This gives us for item (4), $\frac{128}{6} = 21.3$ or, say, 21, whence $T_s = 126$.

Items (5) *to* (13). The secondary current is $I_s = \frac{1,500,000}{6600} = 227$ amperes. From Art. 20, we select $\Delta = 1600$ as a reasonable value for the current density, giving $\frac{227}{1600} = 0.142$ sq. in. for the cross-section of the secondary conductor.

In order to decide upon a suitable width of copper in the secondary coils, it will be desirable to estimate the total space required for the windings so that the proportions of the " window " may be such as have been found satisfactory in practice. The space factor (Art. 41) is not likely to be better than o.1, which gives for the area of the " window " $\frac{2 \times 126 \times 0.142}{0.1} = 358$ sq. in. Also, if a reasonable assumption is that $H = 2.5$ times D (see Fig. 49, page 147), it follows that $2.5D \times D = 358$; whence $D = 12$ inches.

The clearance between copper and iron under oil, for a working pressure of 6600 volts (Formula (12), Art. 16), should be about $0.25 + 0.05 \times 6.6 = 0.58$ in. For the insulation between layers, we might have o.02 in. for cotton, and a strip of o.012 in. fullerboard, making a total of $21 \times 0.032 = 0.67$ in. The thickness of each secondary conductor will therefore be about $\dfrac{12 - (0.58 + 0.67 + 0.58)}{21} = 0.485$ in., which gives a width

of $\dfrac{0.142}{0.485} = 0.293$ in. Let us make this 0.3 in., and build up each conductor of three strips 0.16 in. thick, with 0.006 paper between wires (to reduce eddy current loss) and cotton tape outside. Allowing 0.026 in. for the cotton tape, and 0.012 in. for a strip of fullerboard between turns, the total thickness of insulation, measured across the layers, is $21 \times (0.026 + 0.024) = 1.05$ in.

A width of "window" of 12.75 in. (see **Fig. 52**) will accommodate these coils. The current density with this size of copper is

$$\Delta = \frac{227}{3 \times 0.16 \times 0.3} = 1575 \text{ amps. per sq. in.}$$

Items (14) *to* (25). *H.T. Winding.* $T_p = 126 \times \dfrac{880}{66}$
$= 1680$. This may be divided into 16 coils of 95 turns each, and 2 coils of only 80 turns each, which would be placed at the ends of the winding and provided with extra insulation between the end turns (see Art. 14).

According to Formula (13) of Art. 16, the thickness of insulation—consisting of partitions of fullerboard with spaces between for oil circulation—separating the H.T. copper from L.T. coils or grounded iron, should not be less than $0.25 + 0.03 \times 88 = 2$ 89 in. Let us make this clearance 3 in. Then, since the width of opening is 12.75 in., the maximum permissible depth of winding of the primary coils will be $12.75 - 6 = 6.75$ in. The primary current (Item 17) is $I_p = \dfrac{1,500,000}{88,000} = 17.05$ amps. (approx.). The cross-section of each wire is $\dfrac{17.05}{1600}$ $= 0.01065$ sq. in. Allowing 0.016 in. for the total increase of thickness due to the cotton insulation, and 0.012

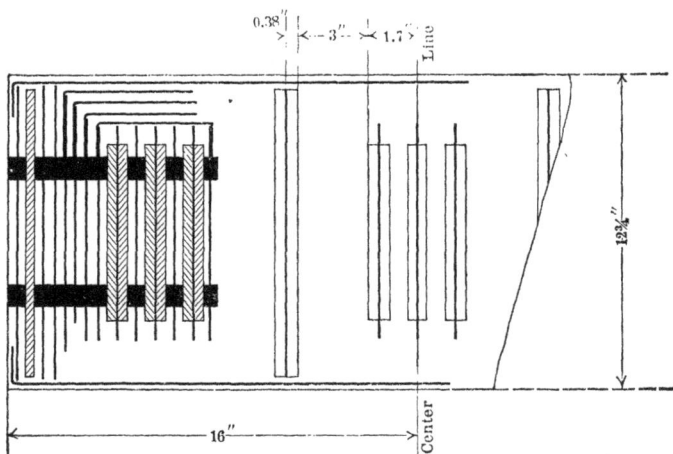

FIG. 52.—Section through Windings and Insulation.

in. for a strip of fullerboard between turns, the thickness of the copper strip (assuming flat strip to be used) must not exceed $\left(\dfrac{6.75}{95}\right) - 0.028 = 0.043$ in., which makes the width of copper strip equal to $\dfrac{0.01065}{0.043} = 0.248$ in. Try copper strip $0.26 \times 0.04 = 0.0104$ sq. in., making $\Delta = 1640$.

The two end coils, with fewer turns, would be built up to about the same depth as the other coils by putting increasing thicknesses of insulation between the end turns. Thus, since there is a total thickness of copper equal to $0.04 \times (95 - 80) = 0.6$ in. to be replaced by insulation, we might gradually increase the thickness of fullerboard between the last eight turns from 0.012 in. to 0.15 in.

Items (26) and (27). Size of Opening for Windings. A drawing to a fairly large scale, showing the cross-section through the coils and insulation, should now be made. Oil ducts not less than $\frac{1}{4}$ in. or $\frac{5}{16}$ in. wide should be provided near the coils to carry off the heat, and the large oil spaces between the H.T. coils and the L.T. coils and iron stampings, should be broken up by partitions of pressboard or other similar insulating material, as indicated roughly in a portion of the sketch, Fig. 52. In this manner the second dimension of the "window" is obtained. This is found to be 32 in., whence the copper space factor is

$$\frac{(1680 \times 0.0104) + (126 \times 0.1445)}{12.75 \times 32} = 0.0875.$$

Items (28) to (41). The Magnetic Circuit. By Formula (1), Art. 2,

$$\Phi = \frac{88,000 \times 10^8}{4.44 \times 50 \times 1680} = 2.36 \times 10^7.$$

Before assuming a flux density for the core, let us calculate the permissible losses.

The full load efficiency being 0.981, the total losses are $\dfrac{1,500,000 \times (1 - 0.981)}{0.981} = 29,000$ watts. Also, since the ratio $\left(\dfrac{W_c}{W_i}\right)$ is approximately 0.925 (see Art. 39, under sub-heading *The Factor b*), it follows that

$$W_i = \frac{29,000}{1.925} = 15,100 \text{ watts,}$$

whence

$$W_c = 29,000 - 15,100 = 13,900 \text{ watts.}$$

Let us assume the width of core under the windings (the dimension L of Fig. 49) to be 11 in. and the width, B, of the return circuit carrying half the flux, to be 5.5 in. Then the average length of the magnetic circuit, measured along the flux lines, will be $2(12.75 + 5.5 + 32 + 5.5) = 111.5$ in.

If the flux density is taken at 13,000 gausses (selected from the approximate values of Art. 20) the cross-section of the iron is $\dfrac{2.36 \times 10^7}{13,000 \times 6.45} = 282$ sq. in. The watts lost per pound (from Fig. 27) are $w = 1.27$, whence the total iron loss is

$$W_i = 1.27 \times 0.28 \times 282 \times 111.5 = 11,200 \text{ watts,}$$

which is considerably less than the permissible loss. It is not advisable to use flux densities much in excess of the selected value of 13,000 gausses for the following reasons:

(a) The distortion of wave shapes when the magnetization is carried beyond the " knee " of the B–H curve.

(*b*) The large value of the exciting current.

(*c*) The difficulty of getting rid of the heat from the surface of the iron when the watts lost per unit volume are considerable.

Let us, therefore, proceed with the design on the basis of 14,000 gausses as an upper limit for the flux density.

If no oil ducts are provided between sections of the stampings, the stacking factor will be about 0.89. A gross length of 27 in. (Item 35) gives 24 in. for the net length, and a cross-section of $24 \times 11 = 264$ sq. in. Whence $B = 13,850$ gausses, and the total weight of iron is $264 \times 111.5 \times 0.28 = 8250$ lb.

The watts per pound, from Fig. 27, are $w = 1.44$, whence $W_i = 11,900$.

Items (42) *to* (49), *Copper Loss.* The mean length per turn of the windings is best obtained by making a drawing such as Fig. 53. This sketch shows a section through the stampings parallel with the plane of the coils. The mean length per turn of the secondary, as measured off the drawing, is 122 in., and since the length per turn of the primary coils will be about the same, this dimension will be used in both cases. Taking the resistivity of the copper at 0.9×10^{-6} ohms per inch cube (see *The Factor k_c*, in Art. 39), the primary resistance (hot) is

$$R_1 = \frac{0.9 \times 122 \times 1680}{10^6 \times 0.0104} = 18.1 \text{ ohms,}$$

whence the losses (Item 44) are

$$(17.05)^2 \times 18.1 = 5260 \text{ watts.}$$

For the secondary winding we have

$$R_2 = \frac{0.9 \times 122 \times 126}{10^6 \times 0.144} = 0.0962 \text{ ohm,}$$

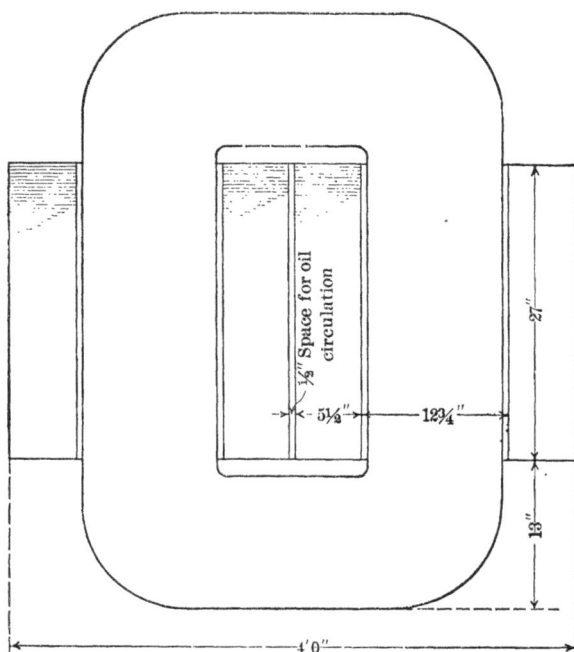

FIG. 53.—Section through Coil and Stampings.

whence the losses (Item 47) are $(227)^2 \times 0.0962 = 4960$ watts, and

$$W_c = 5360 + 4960 = 10,220 \text{ watts,}$$

which is appreciably less than the permissible copper loss.

It is at this stage of the calculations that changes should be made, if desirable, to reduce the cost of materials, by making such modifications as would bring the losses near to the permissible upper limit. The obvious thing to do in this case would consist in increasing the current density in the windings, and perhaps making a small reduction in the number of turns. A considerable saving of copper would thus be effected without necessarily involving any appreciable increase in the weight of the iron stampings. Since this example is being worked through merely for the purpose of illustrating the manner in which fundamental principles of design may be applied in practice, no changes will be made here to the dimensions and quantities already calculated.

The weight of copper (Item 49) is

$$0.32 \left(122 \times 1680 \times 0.0104\right) +$$
$$0.32(122 \times 126 \times 0.144) = 1,700 \text{ lb.}$$

Items (50) *and* (51). *Efficiency.* The full-load efficiency on unity power factor is

$$\frac{1,500,000}{1,500,000 + 11,900 + 10,220} = 0.985.$$

The calculated efficiencies at other loads are:

At $1\frac{1}{4}$ full load................. 0.985
At $\frac{3}{4}$ full load................. 0.984
At $\frac{1}{2}$ full load................. 0.981
At $\frac{1}{4}$ full load................. 0.968

The full-load efficiency on 80 per cent power factor is

$$\frac{1,500,000 \times 0.8}{(1,500,000 \times 0.8) + 22,120} = 0.982.$$

Item (52). *Open-circuit Exciting Current.* Using the curves of Fig. 45 (see Art. 35 for explanation), we obtain for a density $B = 13,850$ the value 23 volt-amperes per pound of core. The weight of iron (Item 40) being 8250 lb., it follows that the exciting current is

$$I_e = \frac{8250 \times 23}{88,000} = 2.15 \text{ amps.}$$

This is 12.6 per cent of the load component, which is rather more than it should be. If the design is altered, as previously suggested, to reduce the amount of copper, this will result in a reduction of the opening in the iron, and, therefore, also of the length of the magnetic circuit. It is, however, clear that the flux density (Item 29) must not be higher than 13,850 gausses. If the design were modified, it might be advisable to reduce this value by slightly increasing the cross-section of the magnetic circuit. The fact that the exciting current component is fairly large relatively to the load current will lead to a small increase in the calculated copper loss (Item 44); but for practical purposes it is unnecessary to make the correction.

Items (53) *to* (56) *Regulation.* Referring to Fig. 52, it is seen that there are six high-low sections, all about equal, since the smaller number of turns in two out of eighteen primary coils is not worth considering in calculations which cannot in any case be expected to yield very accurate results. The quantities for use in Formula (40) of Art. 34 have, therefore, the following values:

$$T_1 = \tfrac{1680}{6} = 280;$$
$$I_1 = 17.05;$$
$$l = 10.15 \times 12 \times 2.54 = 310 \text{ cm.};$$
$$g = 3 \times 2.54 = 7.62 \text{ cm.};$$
$$p = 1.7 \times 2.54 = 4.32 \text{ cm.};$$
$$s = 0.38 \times 2.54 = 0.965 \text{ cm.};$$
$$h = 12.75 \times 2.54 = 32.4 \text{ cm.}$$

whence the induced volts per section are,

$$I_1 X_1 = 475 \text{ volts.}$$

Since there are six sections, and all the turns are in series, the total reactive drop at full load is

$$I_1 X_p = 475 \times 6 = 2850 \text{ volts,}$$

which is only 3.24 per cent of the primary impressed voltage.

By Formula (35) Art. 33, the equivalent primary resistance is

$$R_p = 18.1 + \left(\frac{1680}{126}\right)^2 \times 0.0962 = 35.2 \text{ ohms;}$$

whence

$$I_1 R_p = 600 \text{ volts.}$$

which is 0.683 per cent of the primary impressed voltage.

By Formula (47), Art. 36, when the power factor is unity ($\cos \theta = 1$).

$$\text{Regulation} = 0.683 + 0 = 0.683 \text{ per cent}$$

The more correct value, as obtained from Formula (46) is 0.735.

When the power factor of the load is 80 per cent, the approximate formula—which is quite sufficiently accurate in this case—gives

Regulation $= (0.683 \times 0.8) + (3.24 \times 0.6)$

$\qquad = 2.5$ per cent (approx.) on 80 per cent power factor.

This is very low, and considerably less than the specified limit of 5 per cent.   It is possible that the specified regulation might be obtained with only 4, instead of 6, high-low groups of coils, and in order to produce the cheapest transformer to satisfy the specification, the designer would have to abandon this preliminary design until he had satisfied himself whether or not an alternative design with a different grouping of coils would fulfill the requirements.   It is clear from the inspection of Fig. 52 that an arrangement with only four L.T. coils and (say) sixteen H.T. coils would considerably reduce the size of the opening in the stampings, thus saving materials and, incidentally, reducing the magnetizing current, which is abnormally high in this preliminary design.

Items (57) *to* (61).   *Requirements for Limiting Temperature Rise.*   A plan view of the assembled stampings should be drawn, as in Fig. 54, from which the size of containing tank may be obtained.   In this instance it is seen that a tank of circular section 5 ft. 3 in. diameter will accommodate the transformer.   The height of the tank (see Fig. 55) will now have to be estimated in order to calculate the approximate cooling surface. This height will be about 90 in., and if we assume a smooth surface (no corrugations), the watts that can be dissipated continuously are

$$0.24 \times \left[(\pi \times 63 \times 90) + \frac{\pi (63)^2}{4 \times 2} \right] = 4650;$$

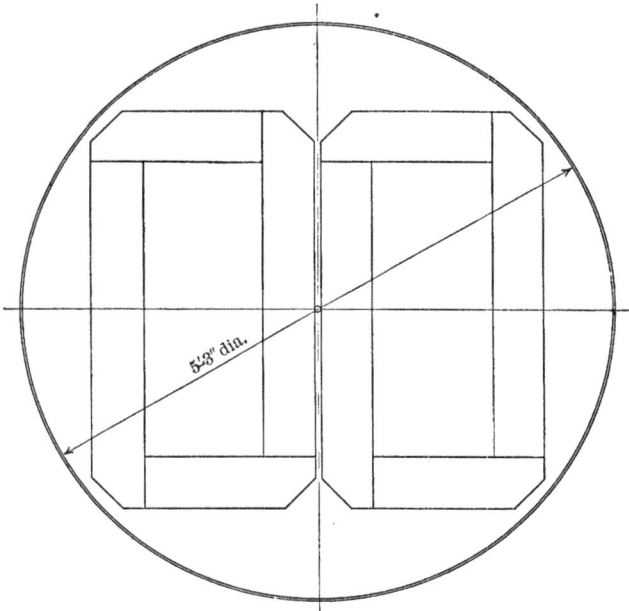

FIG. 54.—Assembled Stampings in Tank of Circular Section.

the multiplier 0.34 being obtained from the curve, Fig. 32 of Art. 25.

The watts to be carried away by the circulating water are $(10,220 + 11,900) - 4650 = 17,470$. From data given in Art. 29, it follows that a coil made of $1\frac{1}{4}$ in.

tube should have a length of $\dfrac{17,470}{12 \times 1.25 \times \pi} = 370$ ft.

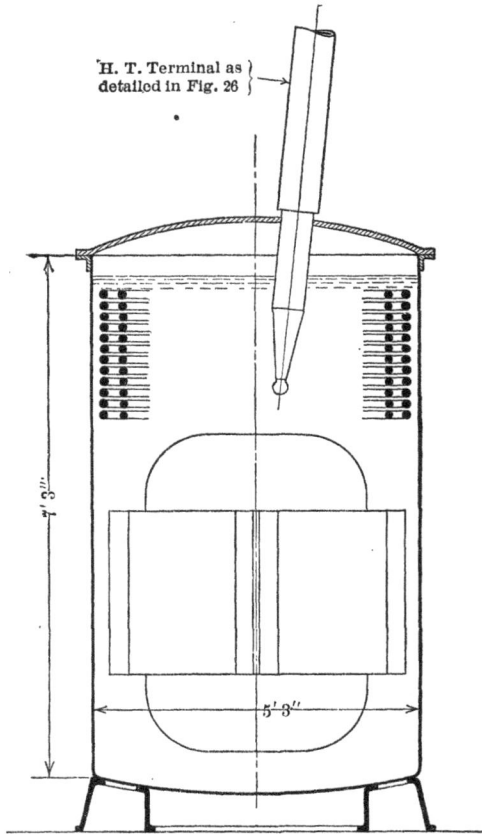

FIG. 55.—Sketch of 1500-k.v.a., 88,000-volt Transformer in Tank.

Assuming the coil to have an average diameter of 4 ft. 8 in., the number of turns required will be about 25.

On the basis of $\frac{1}{4}$ gal. of water per kilowatt, the required rate of flow for an average temperature difference of 15° C. between outgoing and ingoing water is $0.25 \times 17.47 = 4.37$ gal. per minute. This amount may have to be increased unless the pipes are kept clean and free from scale.

The completed sketch, Fig. 55, indicates that a tank 87 in. high will accommodate the transformer and cooling coils, and the corrected cooling surface for use in temperature calculations (see Art. 25) is therefore

$$S = (\pi \times 63 \times 87) + \frac{1}{2}\left(\frac{\pi}{4} \times \overline{63}^2\right) = 18,860 \text{ sq. in.}$$

This new value for Item 57 has been put in the last column of the design sheet; but the items immediately following, which are dependent upon it, have not been corrected because the difference is of no practical importance.

Hottest Spot Temperature. The manner in which the temperature at the center of the coils may be calculated when the surface temperature is known, was explained in Art. 23. It is unnecessary to make the calculation in this instance because the coils are narrow and built up of flat copper strip. There will be no local " hot spots " if adequate ducts for oil circulation are provided around the coils.

Items (62) and (63). Weight of Oil and of Complete Transformer. The weight of an average quality of transformer oil is 53 lb. per cubic foot, from which the total weight of oil is found to be about 7300 lb. The calculated weights of copper in the windings (Item 49) and iron in the core (Item 40) are 1700 lb. and 8250 lb., respectively. The sum of these three figures is 17,250 lb. This, together with an estimated total of 4750 lb. to cover the tank, base and cover, cooling coil, terminals, solid insulation, framework, bolts, and sundries, brings the weight of the finished transformer up to 22,000 lb. (including oil); or $\dfrac{22,000}{1500} = 14.65$ lb. per k.v.a. of rated full-load output.

Several details of construction have not been referred to. It is possible, for instance, that tappings should be provided for adjustment of secondary voltage to compensate for loss of pressure in a long transmission line. These should preferably be provided in a portion of the winding which is always nearly at ground potential. It is not uncommon to provide for a total voltage variation of 10 per cent in four or five steps, which is accomplished by cutting in or out a corresponding number of turns, either on the primary or secondary side, whichever may be the most convenient.

Mechanical Stresses in Coils. The manner in which the projecting ends of flat coils in a shell-type transformer should be clamped together is shown in Fig. 16 of Art. 9. Let us calculate the approximate pressure tending to force the projecting portion of the secondary end coils outward when a dead short-circuit occurs on the transformer. The force in pounds, according to Formula (4), is

$$\frac{T l I_{max}\, B_{am}}{8,896,000}.$$

For the quantities T and l, we have

$$T = T_2 = \frac{T_s}{6} = \frac{126}{6} = 21,$$

and l, being the average length of the portion of a turn projecting beyond the stampings at one end, is

$$l = \frac{10.15 \times 12}{2} - 27 = 34 \text{ in. or } 86 \text{ cms.}$$

The value of the quantities I_{max} and B_{am} depends on the impedance of the transformer. With normal full-load current, the impedance drop is

$$I_1 Z_p = \sqrt{(2850)^2 + (600)^2} = 2910 \text{ volts,}$$

where the quantities under the radical are the items 53 and 54 of the design sheet. In order to choke back the full impressed voltage, the current would have to be about $\frac{88,000}{2910} =$ (say) thirty times the normal full-load value. Thus the current value for use in Formula (4), on the sine wave assumption, will be

$$I_{max} = 30 \times 227 \times \sqrt{2} = 9650 \text{ amperes.}$$

The density of the leakage flux through the coil is less easily calculated; but, since the reactive voltage was calculated on the assumption of flux lines all parallel to the plane of the coil, we may now consider a path one square centimeter in cross-section and of length equal to

the depth of the coil (about 29 cms.) in which the leakage
flux will have the *average* value.

$$B_{am} = \frac{1}{2}\left[\frac{4\pi}{10} \times 21 \times 9650 \times \frac{1}{29}\right] = 4400 \text{ gausses,}$$

whence, by Formula (4),

$$Force \ in \ lb. = \frac{21 \times 86 \times 9650 \times 4400}{8,896,000} = 8640 \text{ lb.}$$

This is the force F of Fig. 16, distributed over the whole
of the exposed surface of the end coil.   An equal force
will tend to deflect outward the secondary coil at the
other end of the stack.   If an arrangement of straps with
two bolts is adopted—as shown in Fig. 16—each bolt
must be able to withstand a maximum load of 4370 lb.
Bolts $\frac{3}{4}$ in. diameter will, therefore, be more than sufficient to prevent displacement of the coils, even on a
dead short circuit.

CHAPTER VI

TRANSFORMERS FOR SPECIAL PURPOSES

44. General Remarks. When applying the fundamental principles of electrical design to special types of apparatus, it is necessary to consider what are the chief characteristics of such apparatus and wherein they differ from those of the more usual types. The apparatus dealt with in the preceding chapters is the potential transformer for use either, as large units, in power stations, or in smaller sizes, as means of distributing electric power in residential or industrial districts. A few special types of transformer will now be considered; but the treatment will be brief, with the object of avoiding useless repetitions. Attention will be given mainly to such distinctive features or peculiarities as may have an important bearing on the design.

45. Transformers for Large Currents and Low Voltages. Electric furnaces are built to take currents up to 35,000 amperes at about 80 volts—usually three-phase. Welding transformers must give large currents at a comparatively low voltage. A current of 2000 amperes at 5 volts would probably be required for rail welding on an electric railroad. Transformers for thawing out frozen water pipes need not necessarily be specially designed because standard distributing transformers—

177

connected to give about 50 volts—are used successfully
for this purpose. A transformer of 12 k.v.a. normal
rating, capable of giving up to 600 amperes with a maximum pressure of 30 volts for short periods of time in
cold weather, will probably answer all requirement for
the thawing of house service pipes up to $1\frac{1}{2}$ in. diameter.
A current of 400 amperes will thaw out a 1-in. pipe in
about half an hour.

In the design of all transformers for large currents,
especially when they are liable to be practically short-
circuited, the leakage reactance (see Art. 34) is a matter
of importance. The permissible maximum current on a
short circuit should be specified. In some cases, separate adjustable reactance coils (usually on the high-
voltage side) are provided for the purpose of regulating
the current from transformers used for welding and similar processes.

Another point to be watched in the design of transformers for large currents is the eddy current loss in the
copper (see Art. 20), which must be minimized by properly arranging and laminating the secondary winding
and leads. The mechanical details in the design of
secondary terminals and leads also require careful attention.

46. Constant-current Transformers. Circuits with
incandescent or arc lamps connected in series require the
amount of current to be approximately constant regard-
less of the number of lamps on the circuit. If it is desired to supply series circuits of this nature from constant
potential mains, special transformers are required, so
designed as to give variable voltage at the secondary

terminals, with a constant voltage across the primary terminals. The variations in the secondary voltage are automatic, being the result of very small changes in the secondary current, brought about by switching lamps in or out of the circuit. In other words, the secondary voltage must follow as nearly as possible the variations in the impedance of the external circuit, so that a doubled impedance would very nearly bring about a doubling of the secondary voltage, the drop in current being as small as possible.

Automatic regulation of this kind may be obtained by means of an ordinary transformer having a large amount of magnetic leakage, as for instance a core type transformer purposely constructed with the primary turns on one limb and the secondary turns on the other limb, as shown diagrammatically in Fig. 1 of Art. 2. The vector diagram of such a transformer has been drawn in Fig. 56, based on the simplified diagram, Fig. 48 (Art. 36), which should be consulted for the meaning of the vectors. The same notation has been used in Fig. 56 as in Fig. 48, and it is to be observed that, on account of the leakage flux being a large percentage of the useful flux, a small reduction in the current, from I_1 to I'_1, will automatically cause the vector E_e (which is a measure of the secondary voltage) to become E_e', just twice as great.

Although by suitably designing a transformer with considerable leakage flux, a small reduction in the reactive drop (the vector I_1X_p of Fig. 56) will produce a large increase in the secondary voltage, it is obvious that still better results would be obtained if the reactance (or amount of leakage flux) could be made to decrease at a greater rate than the current. Thus, if an increase of current could be made to bring about a change in the permeance of the leakage paths, the reactive drop, instead of being proportional to the current, might be made to increase at a greater rate than the current,

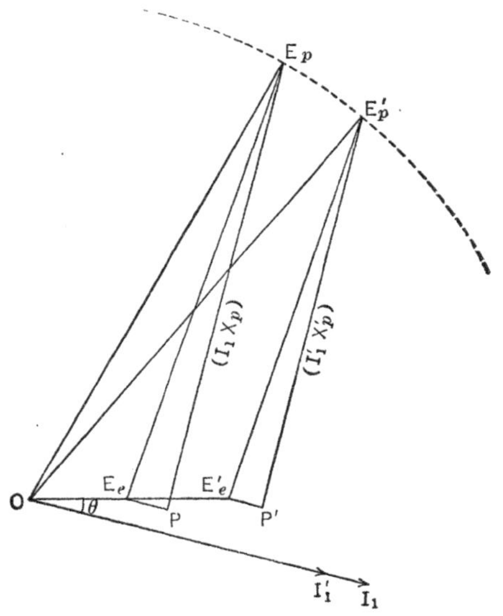

FIG. 56.—Vector Diagram of Transformer with Large Amount of Leakage Flux.

and so bring about the condition illustrated by Fig. 57 where the same result,—i.e., a doubling of the secondary voltage—is seen to be brought about by a very much smaller reduction in the amount of the current.

It is evident that the primary volt-amperes must remain practically constant at all loads, and the fact that the actual secondary output may vary considerably with changes in the resistance of the external circuit, is accounted for by the alteration in the power factor of the primary circuit. Thus, since the input and output of a transformer must be the same except for the internal losses, the changes of input with an almost constant

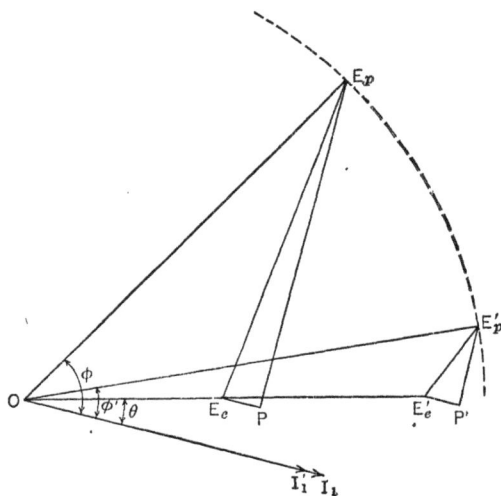

FIG. 57.—Vector Diagram of Transformer with Variable Leakage Reactance.

$E_p I_p$ product are accounted for by the changes in the angle ϕ of Fig. 57.

Fig. 58 illustrates the principle of construction of the constant-current transformer with variable magnetic leakage. One coil is stationary while the other is movable, being suspended from a pivoted arm provided with a counterweight, and free to slide up and down on the cen-

tral core of a shell-type magnetic circuit. The movable
coil may be either the primary or the secondary, and by
careful adjustment of the balance weight, a very small
change in the current may be made to produce a considerable change in the relative position of the coils, thus
greatly altering the relation between the leakage and

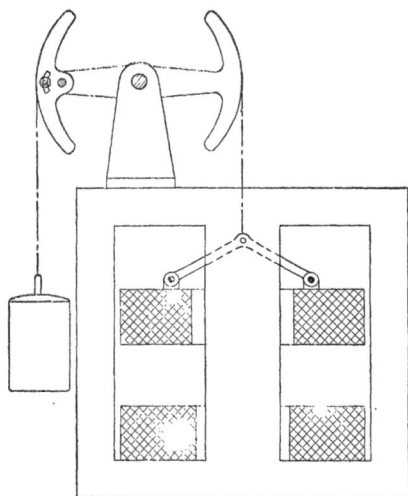

Fig. 58.—Constant Current Transformer with " Floating " Coil.

useful flux components, the (vectorial) sum of which—
passing through the primary coil—must always remain
practically constant.

With the two coils in contact, the maximum secondary
voltage—corresponding to the maximum number of
lamps in series—is obtained; while on short-circuit the
movable coil will be pushed as far away from the sta-

tionary coil as the construction of the transformer will admit. Except for the difficulty of calculating accurately the amount of the leakage flux-linkages corresponding to these two conditions, the design of a constant-current transformer for any given output is a simple matter. Regulation is not usually required over a range greater than from full load to about one-third of full load, and this can be obtained with a current variation not exceeding 1 per cent.

The force tending to move the coils apart can readily be calculated with the aid of Formula (4) Art. 9; but since the quantity B_{am} cannot be predetermined with great accuracy—except in the case of standard designs for which data have been accumulated—final adjustments must be made after completion, by the proper setting of the counterweight.

Constant-current transformers for arc-lamp circuits off constant pressure mains require a secondary current between 6.5 and 10 amperes, and they usually operate in conjunction with a mercury arc rectifier to change the alternating current into a continuous current. Transformers for small outputs may be air cooled, while the larger units should be oil-immersed and, if necessary, cooled by circulating water.

The full-load efficiency of constant-current transformers with movable coils for use on 2200-volt circuits ranges from 90 per cent for 3 kw. output on 60-cycle circuits to 96 per cent for 30 kw. output on 25-cycle circuits.

47. Current Transformers for Use with Measuring Instruments. These transformers are of comparatively

small size, their chief function being to provide a current
for measuring-instruments which shall be as nearly as
possible proportional to the line current passing through
the primary coils. By their use it is possible to trans-
form very large currents to a current of a few amperes
which may conveniently be carried to instruments of
standard construction mounted on the switchboard
panels or in any convenient position preferably not very
far removed from the primary circuit. Again, in the
case of high-potential circuits, even if the reduction of
current is not of great importance, the fact that the secondary circuit of the current transformer can be at
ground potential renders unnecessary the special instruments and costly methods of insulation that would be
required if the line current of high-voltage systems were
taken through the measuring instruments.

A current transformer does not differ fundamentally
from a potential transformer; but since the primary
coils are in series with the primary circuit, the voltage
across the terminals will depend upon the induced
volts, which, in their turn, depend upon the impedance of the secondary circuit. With the secondary
short-circuited, the voltage absorbed will be a minimum, and the input of the transformer will be approximately equal to the copper losses, because a very small
amount of flux will then be sufficient to generate the
required voltage, and the iron losses will be negligible.

The vector diagram for a series transformer does not
differ from that of a shunt transformer, but Figs. 59
and 60 have been drawn to show clearly the influence
of the magnetizing current on the relation between

the total primary and secondary currents. Fig. 59
shows the vector relations when the power factor is
unity, while in Fig. 60 there is an appreciable lag between the current and e.m.f. in the secondary circuit.

When a current transformer is used in connection
with an ammeter only, the essential condition to be
fulfilled is that the ratio $\dfrac{I_s}{I_p}\left(\text{or } \dfrac{I_1}{I_p}\right)$ be as nearly
constant as possible over the whole range of current
values. When the secondary current is passed through

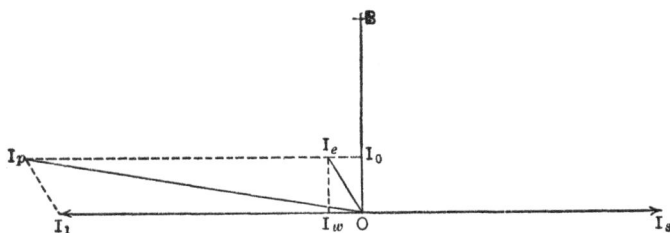

FIG. 59.

the series coil of a wattmeter, it is equally important
that I_s be as nearly as possible opposite in phase to
I_p, or, in other words, that the angle I_pOI_1 be very small.

A diagram, such as Fig. 60, may be constructed for
any given condition of load, the amount of the flux
B—and therefore the exciting current I_e—being dependent upon the impedance of the secondary circuit,
since this determines the necessary secondary voltage.

On the sine wave assumption, it is an easy matter
to express the quantity OI_p in terms of the secondary

current and the two components of the exciting current. The vector OI_1 is a measure of the secondary current, being simply $I_s\left(\dfrac{T_s}{T_p}\right)$, and it is easily seen that

$$I_p = \sqrt{(I_1 \sin\theta + I_0)^2 + (I_1 \cos\theta + I_w)^2},$$

whence the ratio $\dfrac{I_1}{I_p}$ can be calculated for any power factor $(\cos\theta)$ and any values of the secondary current

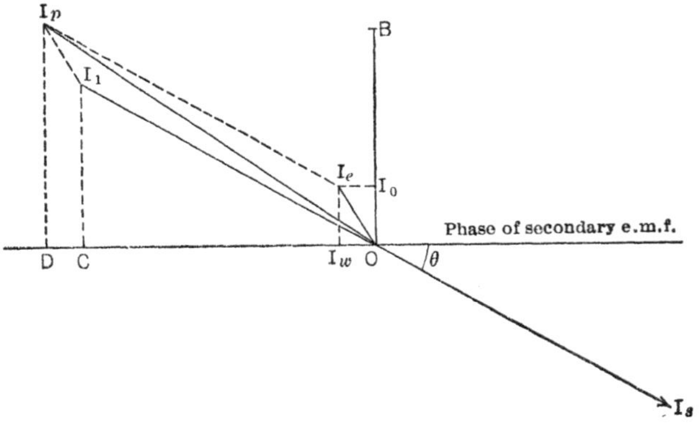

FIG. 60.

and voltage. It is interesting to note that, on a load of unity power factor $(\cos\theta = 1)$, the magnetizing component of the total exciting current does not appreciably affect the relation between the *magnitudes* of the primary and secondary currents, and for all practical purposes the difference, under this particular load condition, is

$$I_p - I_1 = I_w = \frac{\text{iron loss (watts)}}{\text{e.m.f. induced in primary (volts)}}.$$

If this difference were always proportional to the primary current, there would be no particular advantage in keeping it very small; but since the power factor is not always unity, and variations in current magnitudes may be brought about by phase differences, it is always advisable to aim at obtaining an exciting current which shall be a very small percentage of the total primary current.

The phase difference between I_p and I_1 (see Figs. 59 and 60) may be expressed as

$$\text{Angle } I_p O I_1 = \tan^{-1}\left(\frac{I_1 \sin \theta + I_0}{I_1 \cos \theta + I_w}\right) - \theta.$$

This angle must be very small, especially when the transformer is for use with a wattmeter. It should never exceed 1 minute, and should preferably be within thirty seconds. This condition can only be satisfied, with varying values of θ, by making the exciting current (especially the magnetizing component I_0) very small relatively to the main current. It is therefore necessary to use low flux densities in the cores of series transformers for use with instruments, and this incidentally leads to small core losses and a small " energy " component (I_w) of the total exciting current. Flux densities ranging from 1500 to 2500 gausses at full load are not uncommon in well-designed series instrument transformers. Fig. 61 gives approximate losses per pound of transformer iron for these low densities which are not included in the curves of Fig. 27. Although curves for alloyed steel are not given, the losses may be approximately estimated by referring to Fig. 27 (Art. 20) and noting the relative positions of the curves for the two qualities of material.

When the primary current is large, a convenient form of current transformer is one with a single turn of primary, that is to say, a straight bar or cable passing through the opening in the iron core. This is quite satisfactory for currents of 1000 amperes and upward, and the construction is permissible with currents as low as 300 amperes, especially when the transformer is to be used in connection with a single ammeter, i.e., without a wattmeter, or second instrument, or relay coil, in series. The designer should, however, aim to get 1000 to 1500 ampere turns, or more, in each winding of a series instrument transformer.

Although the presence of the exciting current component of an iron-cored transformer renders a constant ratio of current transformation theoretically unattainable over the whole range of current values, this does not mean that any desired ratio of transformation cannot be obtained for a particular value of the primary current. It is, of course, a simple matter to eliminate the error due to the presence of the exciting current by so modifying the ratio of turns $\left(\dfrac{T_p}{T_s}\right)$ that any desired current transformation may be obtained *for a*

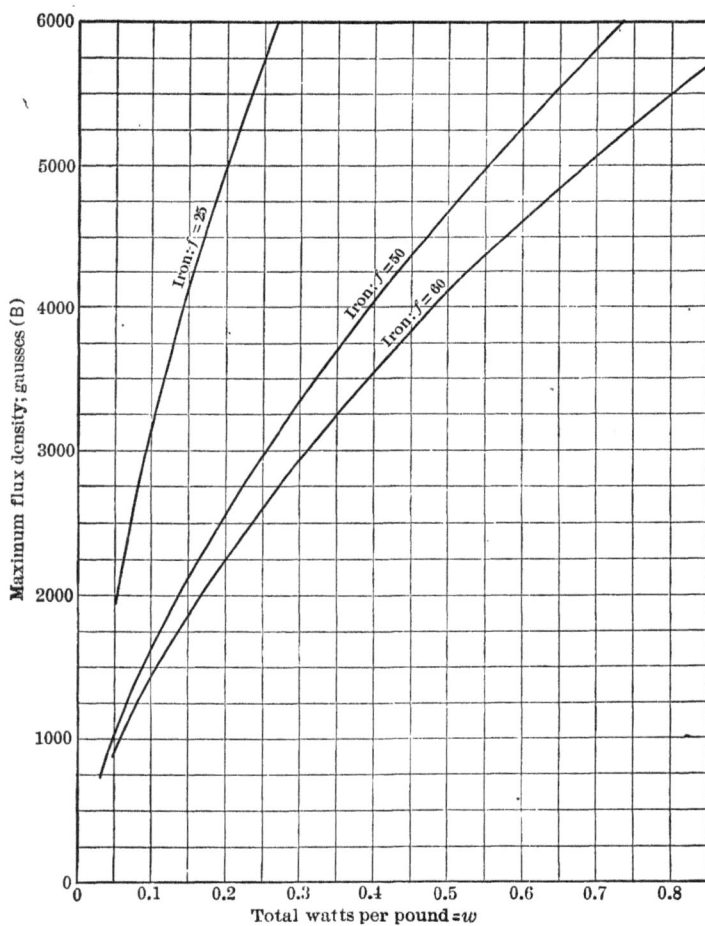

FIG. 61.—Losses in Transformer Iron at Low Flux Densities.

specified value of the primary current. If the ratio of
transformation is correct at full load, it will be practically correct over the range from $\frac{3}{4}$ to full-load current, the error being most noticeable with the smaller
values of the main current. The following figures are
typical of the manner in which the transformation
ratio of series instrument transformers is likely to vary.

Percentage of Full-load Current.	PERCENTAGE DEPARTURE FROM FULL-LOAD RATIO.		
	A	B	C
100........................	o	o	o
75........................	0.25	0.04	0.5
50........................	1.0	0.16	2.0
25........................	3.0	0.5	6.0
10........................	6.0	1.0	12.0

Column *A* gives average values: column *B* shows
how small the error may be in well-designed transformers
for use with wattmeters or other instruments demand-
ing constancy in the current ratios; while *C* refers to
commercial current transformers for use with relays,
trip coils of switches, and other apparatus which does
not call for great accuracy in the transforming ratio.
In all cases a fairly low power factor is assumed, and a
rated full-load output of about 50 volt-amperes. If
the same transformers were to operate on an external
circuit of reduced resistance and unity power factor,
the percentage error would be considerably smaller.

No special features other than reliability of insulation,
and freedom from overheating have to be considered

in connection with series transformers used for operating regulating devices or protective apparatus such
as trip coils on automatic overload circuit-breakers.
The flux density in the core may then be higher than
in instrument transformers.

48. Auto-transformers.   An ordinary transformer becomes an auto-transformer, or compensator, when the

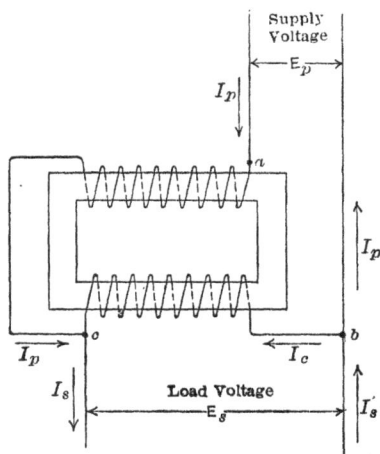

Fig. 62.--Ordinary Transformer Connected as Auto-transformer.

connections are made as in Fig. 62.   One terminal is
then common to both circuits, the supply voltage
being across all the turns of both windings in series,
while the secondary or load voltage is taken off a portion only of the total number of turns.   This arrangement would be adopted for stepping down the voltage;
but by interchanging the connections from the supply

circuit and the load, the auto-transformer can be used equally well for stepping up the voltage.

There is little advantage to be gained by using auto-transformers when the ratio of transformation is large; but for small percentage differences between the supply and load voltages, considerable economy is effected by using an auto-transformer in place of the usual type with two distinct windings.

Let T_p = the number of turns between terminals a and c (Fig. 62);

T_s = the number of turns between terminals c and b;

then $(T_p + T_s)$ = the number of turns between terminals a and b.

The meaning of other symbols is indicated on Fig. 62.

The ratio of transformation is

$$\frac{E_p}{E_s} = \frac{T_p + T_s}{T_s} = r. \quad \cdots \quad (54)$$

If used as an ordinary transformer, the transforming ratio would be

$$\frac{T_p}{T_s} = r - 1. \quad \cdots \cdots \quad (55)$$

The ratio of currents is

$$\frac{I_s}{I_p} = \frac{E_p}{E_s} = r, \quad \cdots \cdots \quad (56)$$

while the current I_c in the portion of the winding common to both primary and secondary is obtained from the equation

$$I_c T_s = I_p T_p,$$

whence

$$I_c = I_p(r - 1), \quad \cdots \quad (57)$$

or, in terms of the secondary current,

$$I_c = I_s\left(\frac{r-1}{r}\right). \quad \cdots \quad (58)$$

None of the above expressions takes account of the exciting current and internal losses.

The volt-ampere output, as an auto-transformer, is $E_s I_s$; but part of the energy passes directly from .the primary into the secondary circuit. For the purpose of determining the size of an auto-transformer, we require to know its equivalent transformer rating. The volt-amperes actually transformed are $E_s I_c$, whence

$$\frac{\text{Output as ordinary transformer}}{\text{Output as auto-transformer}} = \frac{I_c}{I_s} = \frac{r-1}{r}, \quad (59)$$

which shows clearly that it is only when the ratio of voltage transformation (r) is small that an appreciable saving in cost can be effected by using an auto-transformer.

The ratio of turns, and the amount of the currents to be carried by the two portions of the winding having been determined by means of the preceding formulas, the design may be carried out exactly as for an ordi-

nary potential transformer, attention being paid to
the voltage to ground, which may not be the same in
the auto-transformer as in an ordinary transformer for
use under the same conditions. Auto-transformers are,
however, rarely used on high voltage circuits, although
there appears to be no objection to their use on grounded
systems.

Effect of the Exciting Current in Auto-transformers.
In the foregoing discussions, the effect of the exciting
current was considered negligible. This assumption is

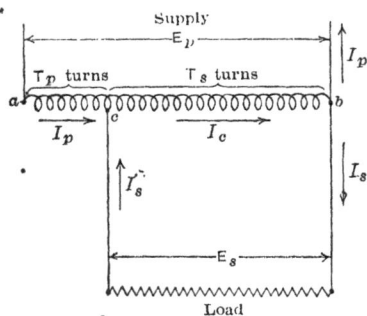

FIG. 63.—Diagram of Connections of Auto-transformer.

usually permissible in practice; but since it may sometimes be necessary to investigate the effect of the
exciting current components, a means of drawing the
vector diagram showing the correct relation of the
current components will now be explained.

Fig. 63 is similar to Fig. 62 except that it shows
the connections in a simplified manner. The arrows
indicate what we shall consider the positive directions
of the various currents.

The fundamental condition to be satisfied is that the (vectorial) addition of all currents flowing to or from the junction c or b shall be zero.   Whence,

$$I_p + I_s = I_c. \qquad \ldots \ldots \quad (60)$$

Let I_e stand for the exciting current when there is no current flowing in the secondary circuit. This is readily calculated exactly as for an ordinary transformer with E_p volts across $(T_p + T_s)$ turns of winding. Then, since the resultant exciting ampere turns must always be approximately $(T_p + T_s)I_e$, the condition to be satisfied under load is

$$I_p T_p + I_c T_s = I_e(T_p + T_s), \quad \ldots \quad (61)$$

which, if we divide by T_s, becomes

$$(r-1)I_p + I_c = rI_e. \qquad \ldots \ldots \quad (62)$$

If I_c in this equation is replaced by its equivalent value in terms of the other current components, as given by Equation (60), we get

$$rI_p = rI_e - I_s. \qquad \ldots \ldots \quad (63)$$

The vector diagram Fig. 64 satisfies these conditions; the construction being as follows:

Draw OB and OE_s to represent respectively the phase of the magnetic flux and induced voltage.   Draw OI_s to represent the current in the secondary circuit in its proper phase relation to E_s.   Now calculate the exciting current I_e on the assumption that it flows through all the turns $(T_p + T_s)$, and draw OM, equal to rI_e, in its proper phase relation to OB. Join MI_s and determine the point C by making $CI_s = \dfrac{MI_s}{r}$. Then, since I_sM is the vectorial difference between rI_e and I_s, it follows from Equation (63) that it is equal to

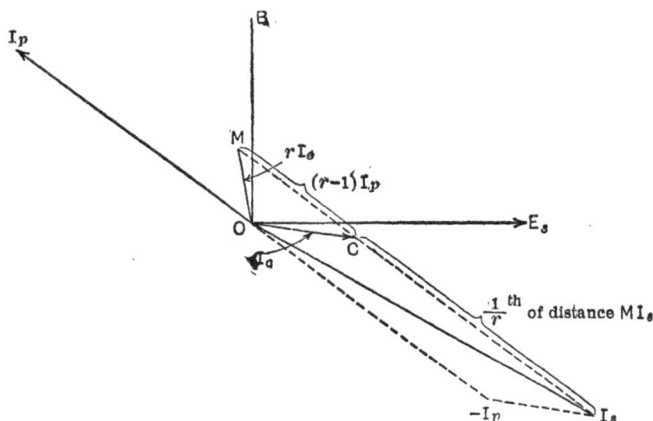

FIG. 64.—Vector Diagram of Auto-transformer, Taking Account of Exciting Current.

rI_p, whence $CI_s = -I_p$, and $CM = (r-1)I_p$. Also, since OC is the vectorial sum of I_s and I_p, it follows from Equation (60) that OC is the vector of the current I_e in the portion of the winding common to both circuits. In this manner the correct value and phase relations of the currents I_p and I_e, in the sections ac and cb of the winding, can be calculated for any given load conditions.

49. Induction Regulators. In order to obtain a variable ratio of voltage transformation, it is necessary either to alter the ratio of turns by cutting in or out sections of one of the windings, or to alter the effective flux-linkages by causing more or less of the total flux linking with the primary to link with the secondary.

The principle of variable ratio transformers of the moving iron type is illustrated by the section shown in Fig. 65. This is a diagrammatic representation of a single-phase induction regulator with the primary coils on a cylindrical iron core capable of rotation through an angle of 90 degrees. The secondary coils are in slots in the stationary portion of the iron circuit. The dotted lines show the general direction of the magnetic flux when the primary is in the position corresponding to maximum secondary voltage. As the movable core is rotated either to the right or left, the secondary voltage will decrease until, when the axis AB occupies the position CD, the flux lines linking with the secondary generate equal but opposite e.m.f.s in symmetrically placed secondary coils, with the result that the secondary terminal voltage falls to zero. If current is flowing through the secondary winding—as will be the case when the transformer is connected up as a " booster " or feeder regulator—the reactive voltage due to flux lines set up by the secondary current and passing through the movable core in the general direction CD, will be considerable unless a short-circuited winding of about the same cross-section as these condary is provided as indicated in Fig. 65.

It is immaterial whether the winding on the movable

core be the primary or secondary; but if the primary is on the stationary ring, the short-circuited coils must also be on the ring.

The chief difficulty in the design of induction regulators

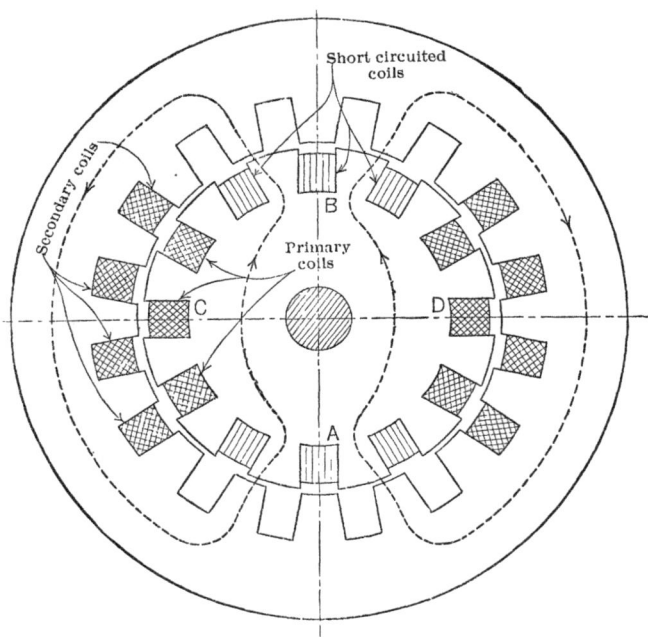

FIG. 65.—Diagram of Single-phase Variable-ratio Transformer of the Moving-iron Type.

arises from the introduction of necessary clearance gaps in the magnetic circuit, and the impossibility of arranging the coils as satisfactorily as in an ordinary static transformer so as to avoid excessive magnetic leakage. A large exciting current component and an appreciable reactive voltage drop are characteristic of the induction voltage-regulator.

Fig. 66 is a diagram showing a single-phase regulating transformer of the type illustrated in Fig. 65 connected as a feeder regulator, the secondary being in series with one of the cables leaving a generating station to supply an outlying district. The movement of the iron core can be accomplished either by hand, or automatically by means of a small motor which is made to rotate in either direction through a simple device actuated by potential coils or relays.

The lower diagram of Fig. 66 shows the core carrying the primary winding in the position which brings the voltage generated in the ring winding to zero. The flux lines shown in the diagram are those produced by the magnetizing current in the primary winding; but there are other flux lines—not shown in the diagram—which are due to the current in the ring winding. It is true that the movable core carries a short-circuited winding—not shown in Fig. 66—which greatly reduces the amount of this secondary leakage flux; but it will nevertheless be considerable, and the secondary reactive voltage drop is likely to be excessive, especially if the ring winding consists of a large number of turns. An improvement suggested by the writer at the time * when this type of apparatus was in the early stages of its development, consists in putting approximately half the secondary winding on the portion of the magnetic circuit which carries the primary winding, the balance

* The year 1895.

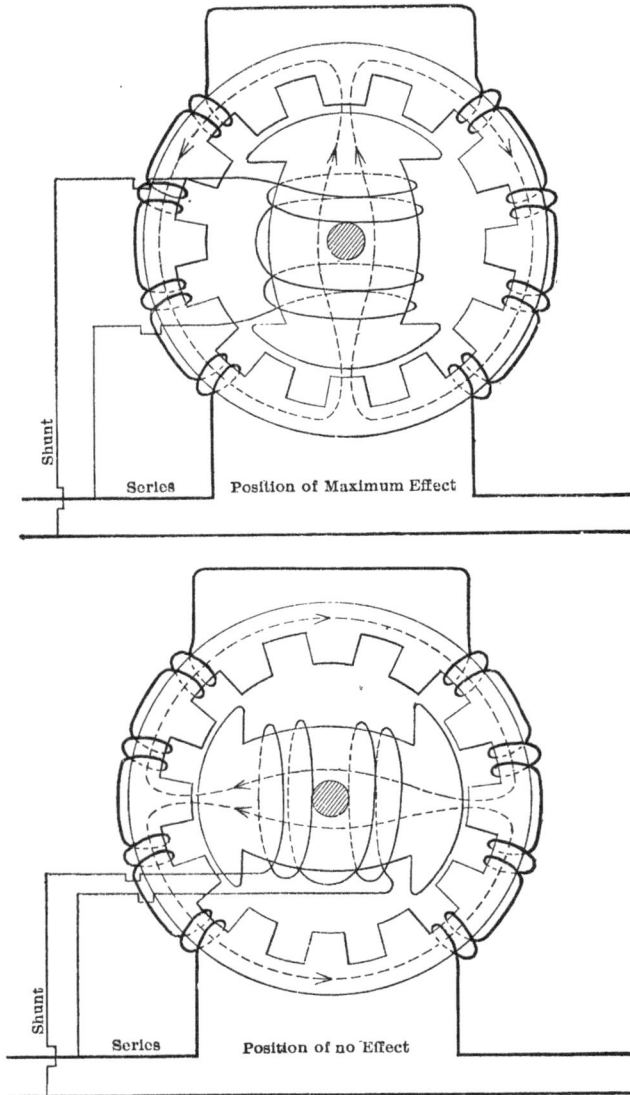

FIG. 66.—Variable-ratio Transformer Connected as Feeder Regulator.

of the secondary turns being put on the other portion of the magnetic circuit. The connections are made as in Fig. 67, the result being that the movement of the rotating core, to produce the full range of secondary voltage from zero to the desired maximum, is now 180° instead of 90° as in Fig. 66; but since, under the same conditions of operation, the ring winding for a given section of iron will carry only half the number of turns that would be necessary with the ordinary type (Fig. 66), the secondary reactive voltage drop is very nearly halved. This is one of the special features of the regulating transformers manufactured by Messrs. Switchgear & Cowans, Ltd., of Manchester, England.

Consider the case of a single-phase system with 2200 volts on the bus bars in the generating station. The voltage drop in a long outgoing feeder may be such as to require the addition of 200 volts at full load in order to maintain the proper pressure at the distant end. If this feeder carries 100 amperes at full load, the necessary capacity of a boosting transformer of the type shown diagrammatically in Fig. 67 is 20 k.v.a. This variable-ratio transformer, with its primary across the 2200-volt supply, and its secondary in series with the outgoing feeder, will be capable of *adding* any voltage between 0 and 200 to the bus-bar voltage. As an alternative, the supply voltage at the generating station end of this feeder may be permanently raised to 2300 volts by providing a fixed-ratio static transformer external to the variable-ratio induction regulator and connected with its secondary in series with the feeder. An induction regulator of the ordinary type (Fig. 66)

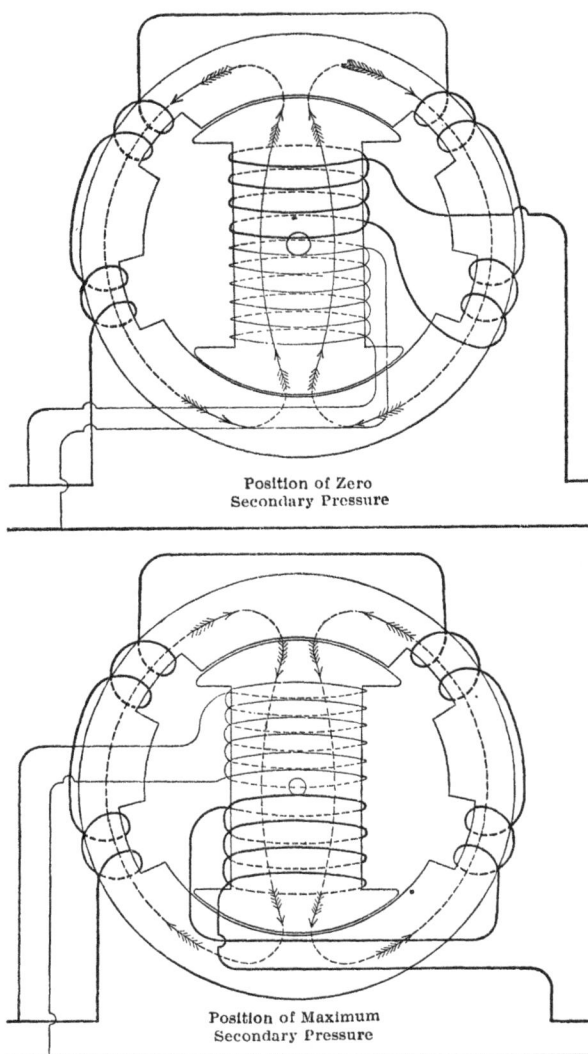

Position of Zero
Secondary Pressure

Position of Maximum
Secondary Pressure

FIG. 67.—Moving-iron Type of Feeder Regulator with Specially Drranged
Secondary Winding.

capable of both *increasing* and *decreasing* the pressure by 100 volts, will then provide the desired regulation between 2200 and 2400 volts. The equivalent transformer output of this regulator will be $\dfrac{100 \times 100}{1000} = 10$ k.v.a.

The Polyphase Induction Regulator. Two or three single-phase regulators of the type illustrated in Fig. 65 may be used for the regulation of three-phase circuits; but a three-phase regulator is generally preferable. The three-phase regulator of the inductor type is essentially a polyphase motor with coil-wound—not squirrel-cage—rotor, which is not free to rotate, but can be moved through the required angle by mechanical gearing operated in the same manner as the single-phase regulator. The rotating field due to the currents in the stator coils induces in the rotor coils e.m.f.'s of which the magnitude is constant, since it depends upon the ratio of turns, but of which the phase relation to the primary e.m.f. depends upon the *position* of the rotor coils relatively to the stator coils. When connected as a voltage regulator for a three-phase feeder, the vectorial sum of the secondary and primary volts of a three-phase induction regulator will depend upon the angular displacement of the secondary coils relatively to the corresponding primary coils.

Mr. G. H. Eardley-Wilmot * has pointed out certain advantages resulting from the use of two three-phase induction regulators with secondaries connected in series, for the regulation of a three-phase feeder. By making

* *The Electrician,* Feb. 19, 1915, Vol. 74, page 660.

the connections so that the magnetic fields in the two regulators rotate in opposite directions, the resultant secondary voltage will be in phase with the primary voltage. The torque of one regulator can be made to balance that of the other, thus greatly reducing the power necessary to operate the controlling mechanism.

INDEX

A

205

PAGE

L

M

O

P

Q

R

S

T

V

SUPPLEMENTARY INDEX

OF TABLES, CURVES, AND FORMULAS

NUMERICAL EXAMPLES